Assessment of Wel
Problems in rel
among PLHI

M. SATYAVALLI

TABLE OF CONTENTS

CHAPTER - I

INTRODUCTION

Background of the Study

Health, disease and illness are universal facts of life. Health is a state of complete physical, mental, social and spiritual wellbeing and not merely an absence of disease or infirmity [WHO, 1947]. Health is multidimensional. Physical health implies the notion of perfect functioning of the body. Mental health refers to emotional, psychological and social wellbeing. Mental health is just as important as physical health (park, 2007). About 14% of the global burden of disease has been attributed to neuropsychiatric disorders. Many health conditions increase the risk of mental disorders. It is important to know that Human Immunodeficiency Virus (HIV) can affect mental health. Some opportunistic infections (which occur when immune system is damaged by HIV) can affect nervous system and can lead to changes in behavior and functioning.

HIV / AIDS has been called a disease of modernity (Schoepf, 2001) and a disease of globalization (Barnett and Whiteside, 2002). The human immunodeficiency virus (HIV) and acquired immunodeficiency syndrome (AIDS) have caused unspeakable distress to millions of people throughout the world.

According to WHO (2012) there are 35.3 millions of people living with HIV / AIDS at globally. As per NACO (2012) there are 2.1 millions of people living with HIV and AIDS. According to Andhra Pradesh AIDS Control Society (2012) there are 5 lakh people living with HIV / AIDS in United Andhra Pradesh.

What is HIV/AIDS

HIV/AIDS is caused by human immunodeficiency virus and is a chronic as well as fatal disease. HIV develops into Acquired immunodeficiency syndrome (AIDS). AIDS is deficiency in the body's defense mechanism or immune system. HIV is transmitted from one person to another through the routes of sexual contact, sharing infected needles for injecting drug use, receiving infected blood, blood products, or organs can allow transmission of HIV. Mothers may transmit the virus to their babies during pregnancy, delivery or breast feeding. Getting a deep needle stick injury from a person infected by HIV during an occupational exposure or reusing infected instruments (needles or surgical instruments) in the hospital setting could increase the risk of transmission. 80% of HIV cases are infected through sexual contact. A person once infected with virus, he or she is infected forever. HIV is a preventable and manageable but not curable. HIV once enters into the body; it uses CD4 cells for replication. T Lymphocytes are processed and stored in the thymus. The most important T lymphocytes are CD4 type and CD8 type. The HIV is especially attracted to the CD4 type of lymphocytes. It enters the CD4 lymphocytes and multiplies in them thereby destroying the CD4 cells. Gradual loss of these CD4 cells results in the loss of immunity in these patients. Immune suppression leads to opportunistic infections.

An opportunistic infection is an infection caused by pathogens (bacterial, viral, fungal or protozoan) that usually do not cause disease in a healthy host, i.e. one with a healthy immune system. A compromised immune system, however presents an opportunity for the pathogen to infect. The HIV itself does not cause death, but the opportunistic infections that occur because of its effect on the immune system can cause death. Person becomes susceptible to many infections that are caused by organisms that naturally reside in the body such as tuberculosis, candidiasis, pneumocystis carni pneumonia. Presence of major opportunistic infections indicates AIDS.

HIV can be detected by either HIV antibody tests such as HIV rapid tests ELISA and Western blot test. Once HIV enters into the human body, it takes 2 to 12 weeks period to develop antibodies. This is called as a window period. HIV cannot be detected with the help of antibody tests until the window period is completed. HIV can be detected much earlier than this with help of antigen tests like DNA PCR, P24 antigen tests. Once diagnosed as HIV positive, he/she is called as PLHIV (person living with HIV/AIDS). To develop into AIDS it may take 2 to 10 years. Antiretroviral drugs are indicated to increase the CD4cells and to decrease the viral load, thereby protecting the person from developing into AIDS. Anti-retroviral therapy (ART) is a combination of 3 drugs such as zidovidine, lamividine and nivarapine which reduces the ability of HIV virus to replicate, thus increases the body's ability to fight against diseases. ART should be started when a person develops opportunistic infections or CD4 cells reaches below 350 in number. ART therapy should be continued for life long. ART improves the quality of life and increases the life span of PLHIV.

Combined retroviral therapy is particularly challenging because of its demanding dosing schedule, common adverse side effects, and the threat of the relatively rapid development of resistance. The requirement of 95% or better adherence to achieve virological success is not the norm for other diseases, where 80% adherence needs to be facilitated, providers need to adopt a non-judgmental stance regarding the "normalcy" of missing medication doses from time to time and to communicate that to patients. Such a stance encourages patients to be open about problems they have with adherence. Other members of the treatment may have more time to spend with a patient to address the many possible barriers to adherence and to facilitate strategies to improve and maintain optimal adherence. People living with HIV have faced a new situation since the arrival of the antiretroviral treatment. HIV has become a long term condition, which not only affects physical health, but also causes psychological and social problems because of stigma and discrimination. These challenges present many decisions and dilemmas for people living with HIV, which involves complex emotional and behavioral problems.

During specific times in the course of HIV disease, patients are particularly vulnerable to acute distress and depression caused by various stressors such as the news of positive result, the initial onset of physical symptoms, a sudden decline in the number of CD4 cells, initiation of antiretroviral therapy and its side effects, the first opportunistic or the first hospitalization and the consequent financial worries.

People with HIV / AIDS have been victims of stigmatization ever since the epidemic began in the 1980's. People living with HIV infection are often ascribed responsibility for their condition because HIV is contracted from behaviors that are considered avoidable, namely unsafe sex and drug use practices (Herek, 1999). Individuals who contract HIV via injection drug practices or multiple sex partners are blamed more for their HIV infection than people described as contracting HIV by having sex with only one partner (Herek, Capitanio & Widaman, 2003). This chronic life threatening disease not only affects PLHIV's physical health, but also impacts their psychological wellbeing as a result of the stigma that is attached to the disease.

ART is capable of improving survival, reducing the occurrence of HIV related opportunistic infections and improving the patient's quality of life.

Assessment of Mental Health

Emotional and mental health is important issues for people with HIV. People with HIV seem to be more likely to experience a range of emotional and mental health problems. It is important to know that HIV can affect mental health. Some opportunistic infections can affect nervous system and can lead to changes in behavior and functioning.

Particular events such as receiving HIV diagnosis, bereavement, the breakdown of a relationship or dealing with starting HIV treatment or its side effects can affect emotional wellbeing. Major stressors can have a major impact on mental health. A good mental health is the ability to respond to the many varied experiences. It is a state of balance between individual and the surrounding world, a state of harmony between oneself and others. Physical ill health can affect the mental health of a person as body and mind are integrated together. Although mental health is an essential component of health, the scientific foundations of mental health are not yet cleared. Psychologists have mentioned some characteristics as attributes of a mentally healthy person such as a person must be free from internal conflicts, he is well adjusted, he searches for identity, he has a strong sense of self esteem, he knows himself, and his needs and problems, he has good self control, he faces problems and tries to solve them intelligently. One of the keys to good health is a positive mental health. Some of chronic diseases can affect the mental wellbeing of a person. When one is living with HIV, mental health problems can affect the physical health by making it harder for PLHIV to take all their HIV medications on time, it is harder for PLHIV to take advantage of support network, interfering with healthy behaviors, such as getting enough sleep and exercise and avoiding risk behaviors such as having unprotected sex, impairing the ability to cope with the stress of daily life.

People who have been diagnosed with severe mental illnesses have a higher prevalence of HIV than the general population. Individuals with severe mental illnesses are more likely to be HIV positive than the general population. Conversely PLHIV have higher rates of mental disorders. There is evidence that psychological distress can shorten the time from HIV to AIDS particularly among PLHIV. A person can notice a change in himself and people around can also notify certain characteristics such as experiencing panic attacks, no longer finding enjoyment in activities which usually make them happy, withdrawing from social interaction, change in memory functioning, sleeping too much or being unable to sleep, feeling "sad" or "empty" much of the time, feeling guilty, feeling tired all the time. Mental illness has been shown to impact HIV transmission as well as disease progression. Depression, substance used disorders and other severe mental illnesses are associated with transmission risk behaviors such as risky sexual activity and drug injection among HIV negative and HIV positive people. In terms of disease progression, poor mental health in general, and depression and substance used in particular have been shown to negatively affect anti-retroviral therapy adherence, which can cause poorer health outcomes and increased risk of HIV transmission .

Good mental health is essential for successful HIV treatment and for living a happy and healthy life. HIV virus can affect mental health. Antiretroviral therapy can enhance the immune status and thereby decreases the occurrence of opportunistic infections. ART is particularly challenging because of its demanding dosing schedule, common adverse side effects and the relatively rapid development of resistance. ART can enhance the quality of life and psychological wellbeing. Besides ART, maintaining healthy life style such as good nutrition, exercise, control of recreational substance use and alternations in sexual risk behavior must be encouraged. Taking charge by making improvements in the areas can enhance patient's feelings of wellbeing and mastery of their lives.

Assessment of Emotional and Behavioral Problems

Emotional and behavioral problems both result from negative responses to stress. Emotional problems are directed towards the self Emotional problems are those problems in which a person turns the problems inwardly and exhibits them in the form of emotional symptoms such as anxiety, depression, withdrawal, thought problems, attention problems or in the form of psychosomatic disorders. Behavioral problems on the other hand are problems such as aggression, rule breaking and intrusive behavior in which a person turns his problems outwardly and express them in the form of "acting out" behavior. Behavioral problems are easily observed by others. Adults who show signs of these behaviors intrude on the rights of others and often violate the norms of community. Assessing emotional and behavioral problems are important for developing successful intervention.

Depression:

HIV does not directly cause depression. But depression is twice as common in people with HIV as in the general public. Recently Bing and Colleagues(2001) assessed national probability sample of nearly 3,000 adults receiving care for HIV infection and found that more than a third screened positive for clinical depression, the most common disorder identified. Half reported the use of an illicit drug in the past year. Psychiatric illness in the context of HIV infection can contribute to diminished health outcomes, increased substance abuse, poor treatment adherence, increased risky sexual behavior or other maladaptive behavior. It is therefore advisable to screen for the presence of depression and drug and alcohol use and to treat or refer these patients to specialists when problems are suspected. Depression is a separate medical issue that needs to be treated. Research shows that depression can speed up HIV progression to AIDS. Sadness, anxiety and irritability are the symptoms of depression. They show loss of interest in things he used to enjoy, feeling of guilt, worthlessness or hopelessness. There will be weight gain or loss, loss of appetite, sleep problems, fatigue, loss of concentration, low sex drive and also thoughts of suicide. Diagnosing depression can be hard in someone with HIV. Some HIV symptoms and side effects of HIV drugs are the same as those of depression. These include fatigue, low sex drive, little appetite, confusion, nightmares, nervousness and weight loss. But true loss of interest in activities that someone used to enjoy is a sign that a person is depressed. Treatment such as psychotherapy and antidepressants can generally help to overcome depression. Not all depressed patients need antidepressant medication. For those with mild but persistent depression, support groups or counseling may be helpful. Thus structured forms of psychotherapy may alleviate distress or both cognitive behavioral therapy and interpersonal therapy have been shown to alleviate depression in HIV positive patients.

Anxiety:

Anxiety is a common symptom in HIV patients. Anxiety is a feeling of panic or apprehension which is often accompanied by the physical symptoms of sweating, shortness of breath, rapid heartbeat, agitation, nervousness, head ache and panic. In addition to somatic complaints, patients with anxiety disorders often present with fear, worry, insomnia, impaired concentration and memory, diminished appetite, ruminations, compulsive rituals and avoidance of situations that make them anxious. Anxiety can accompany depression. Anxiety is a common symptom in HIV infected patients. When anxiety symptoms are severe or persistent, patients may have an anxiety disorder. These disorders include panic disorder, generalized anxiety disorder, obsessive compulsive disorder and traumatic stress disorder. Patients with limited social support may be particularly susceptible to developing anxiety symptoms. Anxiety is felt by clients during and after receiving an HIV test. During the period of stress anyone will be under a state of anxiety. An individual is able to cope up with the associated symptoms of anxiety. But when the individual is not able to do so and starts developing marked symptoms, he/she may require help. Certain anxiety symptoms can be effectively managed without the use of medication. These are patients who prefer to avoid the use of psychotropic medications. Patients with mild anxiety symptoms that do not interfere with function may respond to supportive or behavioral interventions.

Aggressive Behavior:

Aggressive behavior is behavior that causes physical or emotional harm or threat to others. It can range from verbal abuse to the destruction of a victim's personal property. People with aggressive behavior tend to be irritable, impulsive and restless. Aggressive behavior is intentional, which means that an individual does something purposefully, violates social norms and causes a breakdown in a relationship. Emotional problems are the most common cause of aggressive behavior. Aggression can be reactive relational aggression and instrumental relational aggression. Reactive relational aggression (hostile, affective, retaliatory) is used as a response to feeling attached, threatened or provoked. Usually the person who exhibits this type of aggression feels provoked to do so. Instrumental relational aggression (goal oriented) is used by an individual in order to get what he wants.

Aggressive behavior can be very stressful and it is one type of challenging behavior that can result from dementia. The aggressive behavior may be verbal e.g. Swearing, screaming, shouting, making threats and also can be shown in physical behavior e.g. Hitting, pinching, scratching, hair pulling, biting. Finding ways to prevent and manage the behavior will improve the quality of life. Listening to favorite music may help to reduce aggressive behavior.

Thought Problems:

People living with HIV can sometimes develop thinking and memory problems. These problems can vary from few symptoms to severe problems with memory and concentration. A study under process by national institute of neurological disorders and stroke, U.S (2015) revealed that the natural history of neuro cognitive impairment in human immunodeficiency virus infected individuals remains poorly understood. While the advent of highly active antiretroviral therapy had led to a decreased incidence of the most severe form of HIV associated neuro cognitive disorders, HIV associated dementia, it does not appear to have impacted overall prevalence of HIV associated neuro cognitive disorders. This evidence suggests that the central nervous system could be an important reservoir for HIV regardless of cumulative time on treatment.

It initially was thought that good treatment could prevent all HIV related memory problems. However even people with low HIV viral loads can have these problems. It may be caused by HIV affecting the brain and spinal fluid. Research studies revealed that ART lead to a decreased incidence of the most severe form of HIV associated neuro cognitive disorders and HIV associated dementia. A study prior the widespread introduction of Highly active antiretroviral therapy suggests HIV infected patients who show subtle signs of mental impairment may actually have a very early manifestation of AIDS related dementia. Even relatively minor memory, movement or speaking problems early in the course of disease can be a sign of later dementia. Adherence to the Antiretroviral therapy could decrease the neuro cognitive problems of HIV clients.

Attention Problems:

HIV patients in early stages show mild difficulties in concentration and attention. A study by Melrose et al (2009) showed that HIV in the brain is associated with cognitive impairment. Without combination antiretroviral therapy, cognitive impairment increase with successive stages of HIV. An example of attention problem is the classic complaint of walking into a room and forgetting why one has done so.

Forgetting whether one locked the door after leaving the house and forgetting where one parked the car are other examples of attention related forgetfulness.

HIV infects the brain and leads to decreased efficiency of the brain network. Therefore the neural network adopts by recognizing and by using reverse brain regions to compensate for the efficiency during brain activation. Normally blood brain barrier serves as a protective mechanism by preventing entry of foreign substances; disruption of the blood brain barrier by HIV contributes to the progression of infection. The virus is able to enter the brain through infected cells that pass though the BBB to replace immune cells surrounding the blood supply in the brain. When infected immune cells are able to migrate into tissues compared to uninfected cells. HIV associated dementia is characterized by symptoms of cognitive, motor and behavioral disturbances. There is often progressive slowing of cognitive functions, concentration and attention, memory, new learning, sequencing problem solving and executive control. HIV associated dementia also can present with behavioral which mainly takes the form of apathy, loss of motivation, poor energy, fatigue and social withdrawal. HIV clients display at least two out of the five following cognitive symptoms for more than one month namely impaired attention, slowing in processing information, difficulty with reasoning, impaired memory and impaired speech. They also exhibits one of the following such as acquired abnormality in motor functioning by clinical examination or neuropsychological testing, decline in motivation, emotional control or social behavior. ART may reverse the disease process. Beside ART, cognitive skill building is necessary to improve cognitive functioning.

Somatic Problems:

Unexplained or multiple somatic symptoms are strongly associated with coexisting depressive and anxiety disorders. Routine screening for somatic symptoms is essential because some patients may not consider their symptoms important until significant morbidity is resulted. HIV infected patients who have complaints of fatigue or insomnia should routinely be assessed for major depression (Perkins D. O.1995). HIV patients indicated somatic complaints more frequently (Fukunishi. I et al1997). Among these, the presence or absence of five complaints such as abdominal distress, sleep disturbance, smothering sensations, chest pain or discomfort, numbness or chills was significantly discriminated based on scores on depression. The results suggest that depressive symptoms accompany several somatic complaints in HIV patients without AIDS. Although a longitudinal follow up study is required, the assessment of somatic symptoms may bring clinical benefits such as earlier diagnosis and or intervention for a depressive state. Anti depressants and cognitive behavioral therapy are both effective for treatment of somatic syndromes such as irritable bowel syndrome, fibromyalgia, pain disorders and chronic headache.

HIV patients report improvement after the onset of antiretroviral therapy mainly with regard to physical health, pain symptoms, physical performance and resistance. Mental stress however appears to show only small changes or no changes at all. In this situation, it may be helpful for affected persons to receive psychological support in the form of music.

Intrusive behavior:

Every one experiences intrusive thoughts and doubts, that are usually absurd and are the opposite of what they want to do or think. When irrational thoughts are impossible to stop, it can lead to obsessive compulsive disorder. The difference is that people with OCD have more frequent and distressing thoughts than others because of the meaning they attach to the thoughts and the way they respond to them. Intrusive thoughts are major symptoms for many people who suffer with anxiety disorders among PLHIV. Many people experience intrusive thoughts with anxiety and they may cause significant distress. An intrusive thought is unwelcome thought, image or unpleasant idea that may become an obsession is upsetting or distressing and can feel difficult to manage or eliminate. Many people experience the type of bad or unwanted thoughts that people with more troubling intrusive thought have, but most people can dismiss these thoughts. When intrusive thoughts occur involuntary with obsessive compulsive disorder, patients are less able to ignore the unpleasant thoughts and may pay undue attention to them causing the thoughts to become more frequent and distressing.

Intrusive thought can be as simple as feeling that the hands are contaminated by germs. This intrusion is usually followed by a compulsion to wash the hands repeatedly to clean them. Intrusive thoughts may involve violent obsessions about hurting others or themselves. Sexual obsessions involve intrusive thoughts or images of kissing, touching, fondling, rape etc. Intrusive thoughts may result into depression.

People who are clinically depressed may experience intrusive thoughts more intensely, and view them as evidenced by that they are worthless or sinful people. Treatment for intrusive thoughts is similar to treatment for OCD. Mild cases can also be treated with cognitive behavioral therapy, which helps patients identify and manage to unwanted thoughts. Assessment of this intrusive behavior is important to assist PLHIV in reminding them that these thoughts are automatic and they can safely ignore them.

Rule breaking behavior:

A common feature of the antisocial, rule breaking behavior is the failure to follow moral guidelines. Impairments to the emotional component that comprises the feeling of what is moral is viewed as the primary deficit in antisocial, although some disruption to the cognitive and cognitive emotional components of morality. Externalizing symptoms, especially rule breaking and aggressive behavior can be seen in HIV clients. Aggression and rule breaking are influenced by genetics and environment. It is a conduct disorder. Lying, stealing, truancy are some of the signs of rule breaking behavior. Research studies show that there are HIV infected people who do engage in unprotected sex, just as there are HIV negative people. Recreational drug use also affects safe sex regardless of serostatus [particularly depression] and may compromise sexual functioning. Comprehensive prevention aimed at HIV infected persons should address these various issues and should be an integral part of HIV prevention.

Das. S (2009) studied emotional and behavioral problems among 50 HIV infected children. Higher behavioral problems that is rule breaking among those children whose parents disclosed their HIV status and those children who are discriminated by other children due to their HIV status.

PLHIV mostly they are reluctant to inform the sero status to sexual partner for the fear of rejection. They indulge in unprotected sex with the partners. Maintaining confidentiality without revealing HIV status to their sexual partner, indulging in unsafe sex, unsafe needle sharing by intravenous drug users and substance abuse are seen in PLHIV, which are risky, leads to further transmission of HIV. So the psychologists need to identify the rule breaking in PLHIV in order to expand his sense of individual and moral responsibility of not to harm others. There are two important factors, one is negative nature of the behavior and maintaining the therapeutic process as crucial to the decision making process to help the client to develop insight about their anti social behavior and help them to take correct decisions towards adopting positive behavior. The wider responsibility of psychologists to society in which they should be concerned about and work to mitigate the causes of human suffering.

Withdrawn behavior:

Withdrawn is the action of ceasing to participate in an activity. It is a sign of depression. Social withdrawn could be observed in HIV infected people. Conduct problems in clients with HIV may have dangerous complications of risk behaviors such as intravenous drug use and unprotected sex may result in transmission of the virus. Ramafedi [1998] found that an alarming 83% of HIV positive adolescents surveyed did not consistently use condoms during sexual encounters. Behavioral problems can be seen in PLHIV due to direct effect of virus on central nervous system. Mostly externalizing behavior problems may be associated with environmental factors rather than direct effect of HIV on central nervous system.

PLHIV generally show the withdrawn behavior from the time of initial diagnosis of HIV. It is useful to anticipate and expect a patient to respond to the diagnosis of HIV infection with a wide range of feelings. Expecting them and perhaps alerting the patient to their possible occurrence can be helpful. Empathy goes a long way, as does making the patient understand that he or she is not alone and that may people have gone through and are currently functioning well.

Most people infected with HIV struggle with issues of disclosure of HIV status to others, particularly when first diagnosed. Health care providers should encourage candor between patients and their sexual and needle sharing partners and discuss of safer behaviors in a non-punitive manner, while acknowledging the difficulty in both initiating and maintain certain behavior changes. It is important to help patients resist the desire to either withdraw or isolate themselves.

Withdrawn behavior is observed in PLHIV due to stigma, isolation, multiple losses associated with the disease and frequent sickness. Family stressors are at risk for psychological dysfunction. Additionally, these stressors often exist in the background of environmental strains, such as poverty violence and drug use [Mellins

& Ehrhardt, 1994, Sherwer & Boland 1994]. Physicians and psychologists need to identify the withdrawal behavior symptoms and in order to conduct counseling in resolving their psychological issues.

Understanding the emotional and behavioral problems of PLHIV is critical for improving their mental health and quality of life. Mental health problems are associated with poor medication adherence even in adults. Early identification and management of emotional and behavioral problems helps the PLHIV in adherence to antiretroviral therapy, thereby enhances the quality of life.

Appreciation of Music

Music is one art that has existed in all cultures almost from the time humanity started. It took different steps and developments and now almost all cultures and countries have their own refined musical form. Music is a complex and universal social behavior. It enlightens the soul. Swami prajananda (1965) clearly states that music elevates and animates the levels of consciousness and that music is the gift of god to humanity as it consoles the mind and helps in concentrating and meditating upon universal center. It may be stated that music is a scientific art and an artistic science. According to Bruscia (1998), music therapy is a systematic process of intervention where in the therapist helps the client to promote health using musical experiences.

Music appreciation can include one's interest in music and music aptitude. Music is an important interest of many people. The type of music vary greatly, from pop to classical and jazz to ethnic. Some people just enjoy listening and learning about the types of music they like, their composers and the performers and their history. Many who got great satisfaction from music are not able to, or do not want to, perform themselves. Music interest is part of wider activity including going to concerts, playing an instrument or singing or even composing music.

Musical aptitude refers to a person's innate ability to acquire skills and knowledge required for music activity. Music aptitude defined as, it is the potential to achieve in music (Gordon, 2003). Throughout the 20th century psychologists have designed tests to measure music constructs such as talent, ability, musicality and aptitude. Recent research reveals that music aptitude, like all human characteristics, is normally distributed in the population. All persons have the potential to achieve in music. Relatively few have high aptitude, a similar number have low aptitude. Music aptitude testing helps music teachers meet the unique needs of each student. Research indicates that music aptitude is developmental during the early years of life. A child's aptitude at birth is innate, but can fluctuate until about age nine according to the richness and diversity of musical experiences the child undergo. After age nine one cannot expect to achieve in music beyond the limit of one's stabilized music aptitude. It is essential that children receives lots of high quality informal guidance and formal instruction in music in order to best realize their potential for music fulfillment throughout life. Informal music experiences prior to age five is particularly important.

Music is a highly versatile and dynamic therapeutic modality, lending itself to a variety of music therapy techniques used to benefit both those living with life threatening illnesses and their family members and care givers. HIV infection affects the entire person in his or her environment. Medical treatment address the spreading of the virus in the body and thereby the progression of the illness. Frequently the mental burden involved is neglected. Treatment of an HIV infection should be therefore be interdisciplinary. There are many ways to enjoy music including listening, singing, following the rhythm and moving to the music. Evidence suggests that music can improve someone's mood, behavior and wellbeing. Favorite songs or pieces of music can be powerful prompts for reminiscence. Music therapy can be a supportive form of therapy for adults who are going through an emotional crisis or experiencing mental health problems such as anxiety, depression or psychosis. It can be useful for people who are experiencing distress and isolation arising from illness or disability. Music therapy provides an opportunity to reflect on relationships and explore new ways of relating to others. Engagement in music therapy can create positive changes in emotional wellbeing and helps increase confidence, self awareness, self esteem and the ability to form and sustain relationships.

It is appealing to note that music therapy is a technique that is diverse in cultures, regions and unlike many other techniques, it has been used since ancient times to improve health. Wainberg et al (2007) write that patient with mental illnesses are at the highest risk for HIV in Brazil, and no techniques have been as such established for prevention. They conclude that music therapy CBT as one of the best ways for prevention and treatment of HIV patients in Brazil as their culture thrives on music, dance and festivities. Hanser (2009) confirmed that use of music therapy as part of modern investigative treatment for HIV being beneficial in the same manner as was Ayurvedic medicine in ancient times in Chineese culture.

Lisa M. Gallagher (2002) a music therapist reported in his study that the music helped improvement in mood while pain, anxiety, depression and even shortness of breath decreased among seriously ill patients. His findings prove that music is a "Universal Language" that can have a positive impact on all patients and even their caregivers.

A study reported in Health Day News stated that exposure to music can dramatically improve the mental and physical condition of the patients receiving palliative care.

It is important for the dying, or those with terminal illness that approaches are used which integrate the physical, psychological, social and spiritual dimensions of their being. With music therapy patients become active and no longer feel helplessly exposed to their future lives with a conscious and unconscious focus on HIV. Music therapy helps them to discover not only inner resources and gather strength to accept and endure the disease, but also to discover new resources of life energy and power.

Music acts like magic key as stated by Maria von Trapp. Music is a systematic application of music by therapist. It is meant to bring about helpful changes in the emotional or physical health of the client. Music has the ability to experience an altered state of physical arousal and subsequent mood by processing a progression of musical notes of varying tone, rhythm and instrumentation for a pleasing effect. Biochemical theory states that music is a sensory stimulus that is processed through the sense of hearing. Sound vibrations are chemically changed into nervous impulses that activate either the sympathetic or parasympathetic nervous system. Entertainment theory suggests that oscillations produced by music are received by the human energy field and various physiological systems entrain with or various match the hertz [oscillation] of the music. Metaphysical theory suggests that music is divine in nature. In the context of present day, the film music is taking a more dominant role and is becoming more popular among both the elite as well as the lower classes of the society.

In the present study, the investigator plans to assess the level of wellbeing, emotional and behavioral problems among PLHIV with respect to their interest in music among PLHIV so as to suggest music therapy as an intervention to overcome the psychiatric problems and to improve their wellbeing.

Significance of the Study

There can be no health without mental health. Mental and emotional wellbeing is essential to overall health. Positive mental health helps realise people their full potential, cope with the stress of life, work productively and make meaningful contributions to their communities. Being diagnosed with HIV would cause stress. Mental health problems can affect anybody. Living with long term illness like HIV can lead to many mental health problems.

Infection with HIV is an important global health challenge. In the 1980s infection with HIV was equated with impending death as people acquired immune deficiency syndrome. However the introduction of ART in 1990 transferred the outlook for people diagnosed as HIV positive. These people now face the prospect of living with a chronic disease and having to reframe their lives and relationships. Being HIV positive became less of a threat and the term people living with HIV (PLHIV) came into common use, to reflect the long standing and less risky nature of the problem. Yet, despite the improvement in medical treatment and adoption of preventive strategies, the spread of HIV is continuing globally.

HIV continues to be as a major public health issue. Since 2000, 38.1 million people have become infected with HIV and 25.3 million people died of AIDS related illnesses. In 2014, an estimated 36.9 million people were living with HIV (including 2.6million children) a global HIV-prevalence of 0.8%. The vast majority of this number live in low and middle economic countries. In the same year 1.2 million people died of AIDS related illnesses.25.8 million people living with HIV are in Sub Sahara Africa, accounting for 70% of the global total. As of March 2015, 15 million PLHIV were receiving ART (including 8,23,000 children)- representing 41% of those in need.59% of people living with HIV are still not receiving treatment.

Particular events such as receiving HIV diagnosis, disclosing that diagnosis, bereavement, the breakdown of a relationship, financial worries and work problems, or dealing with starting HIV treatment or its side effects, can result in feelings of deep unhappiness and emotional distress. This can include feeling emotions such as anger, guilt, fear, sadness and loneliness which are difficult to manage and may interfere with their ability to get on with daily life. Support from family and friends can be very helpful at such times. ART is a boon for PLHIV. It not only decreases the physical symptoms, but also increases the quality of life of PLHIV by increasing the CD4 count, so that the person's ability to fight against diseases will be preserved.

PLHIV are more at risk for depression. Unfortunately the side effects of the ART drugs may lead to depression in PLHIV. Mental health and substance use problems are common among patients infected with HIV and may impede adherence to ART regimen. Psychiatric and substance abuse disorders among PLHIV may impair quality of life, adversely affect the need for and use of health services, impact health outcomes and compromise adherence with complicated medical regimens. So it is imperative to help the PLHIV for adherence to ART. Adherence to ART is important for PLHIV because it is a chronic and fatal disease. Mental health and emotional problems like anxiety, depression, aggressive and intrusive behavior may interfere with adherence to ART among PLHIV unless the mental health problems are identified, the quality of life of PLHIV may not be improved as they may not become adherent to ART.

Music therapy can be a supportive form of therapy for adults who are going through emotional crisis or experiencing mental health problems such as anxiety, depression or psychosis. It can be useful for people who are experiencing distress. Gallagher (2002) focused on 200 patients battling several types of cancer, non cancerous tumors, pain disorders, sickle cell disease, AIDS and other life limiting diseases between the ages of 24 and 87. Among them 30% had some sort of musical background. Music therapy first involved the patient choosing the style of music and she wanted to hear. The therapy averaged approximately 25 minutes. Physical and psychological tests were conducted both before music therapy and after it. The results indicated that in addition to marked benefits in patient anxiety, mood, pain and shortness of breath, more than 80% of patients said their mood had improved following music therapy. So the investigator felt the need to assess the mental and emotional wellbeing along with their interest towards music in order to resolve the problems through counseling and also to refer them for appropriate medical treatment besides recommending complimentary health services like music therapy.

Problem Statement:

Assessment of Wellbeing, Emotional and Behavioral Problems in relation to Appreciation of Music among people living with HIV on Anti retroviral Therapy.

Conceptual frame work:

The concepts in the study are wellbeing, emotional and behavioral problems, appreciation of music among PLHIV on Anti retroviral therapy. Appreciation of music is an independent variable that may influence ones' wellbeing and emotional and behavioral problems of PLHIV. Appreciation of music measures both music interest and music aptitude.

Demographic variables consist of age, gender, education, occupation, marital status, religion, income, area of living, caste, number of dependents, living with partner, family support, social support, duration of ART and HIV disclosure status. These attributes which act as independent variables.

Emotional and Behavioral problems include eight syndromes such as Anxiety/depression, withdrawn, somatic complaints, thought problems, attention problems, aggressive behavior, rule breaking behavior, intrusive and other problems. Mental wellbeing also influenced by the appreciation of music besides emotional and behavioral problems. Mental wellbeing consists of anxiety, depression, energy, positive wellbeing, total wellbeing sub scales. The association of above concepts are shown in a schematic representation.

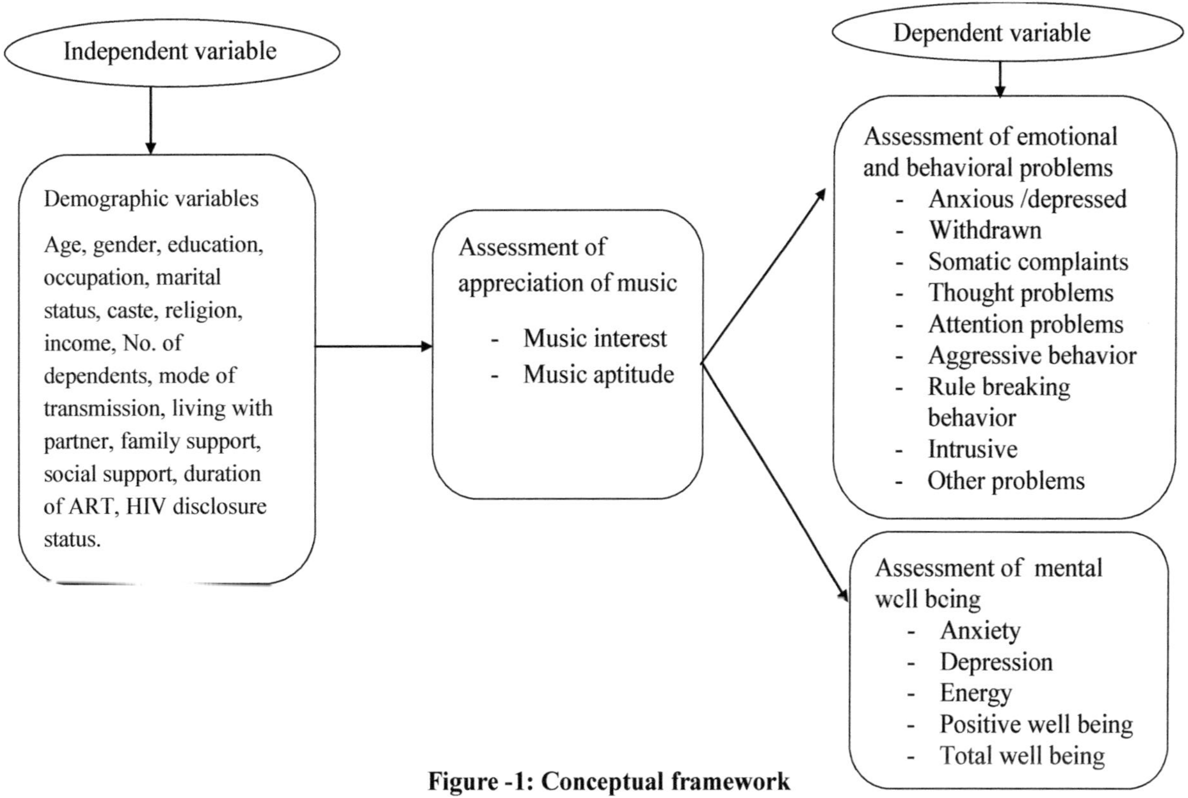

Figure -1: Conceptual framework

CHAPTER - II

REVIEW OF LITERATURE

A systematic review of literature and appraisal of all the relevant scholarly literature on the specific topic has been organized and presented under the following headings:

(I) The literature related to mental wellbeing of PLHIV.

(II) The literature related to emotional and behavioral problems of PLHIV.

(III) The literature related to association between psychological health and antiretroviral therapy.

(IV) The literature related to music therapy.

(V) The literature related to influence of Demographic variables on mental health of PLHIV.

I. Literature related to mental wellbeing of PLHIV.

Heywood & Lyons [2016] conducted a study in Australia on HIV and elevated mental health problems: Diagnostic, treatment and risk patterns for symptoms of depression, anxiety and stress in a national community-Based cohort of gay men living HIV. Their study findings from a national community based sample of 357 HIV positive Australian gay man aged 18 years and older. Participants reported elevated rates of depression, anxiety and generalized stress symptoms. Risk factors for elevated mental health concerns included experiences of internalized stigma and discriminations. Anxiety was also associated with lower CD4 counts. A key protective factor was access to social support. The type of support, in particular emotional support was found to be more important than the source of support.

Charney, Meredith, Keane & Terance (2013) studied the psychological, psycho-social and physical health status of HIV positive refugees. To understand the current status of refugees living with HIV, 84 HIV positive adults living including 26 refugees, 25 Puerto Rican individuals and 33 individuals born in U.S. were interviewed. They assessed psychological, psycho-social and physical domains across these groups to collect comparative information on the status of HIV positive refugees.

Results revealed that HIV positive refugee individuals experienced greater problems in functioning than the other groups with the most significant differences observed in psycho-social functioning. These results underscore the importance of approaching refugee mental health from a multidisciplinary frame work and have implications for the development of interventions for resettled HIV positive refugees.

Gupta, Wyatt, Gail & Honghu (2008) conducted a cross sectional study to characterize similarities and differences in the relationships, psychological wellbeing and sexual behaviors among Indian women (N=459). Both HIV positive (N=216) and negative (N=243) women from urban and rural areas in India were included in this study. Results revealed that HIV positive women were significantly more likely to report marital dissatisfaction, a history of forced sex, domestic violence, depressive symptoms and husband's extra marital sex when compared to the HIV negative women. Findings also indicate that specific factors related to the quality of the marital relationship such as domestic violence, marital dysfunction and depressive symptoms may be related risks for women.

Thomas (2006) studied on how illness and the daily and long term duties of caring amongst a sample of households in the Caprivi region of Namibia impacts upon the physical and psychological wellbeing of ill people and their careers. While optimism and enhanced wellbeing were recorded during periods of illness remission AIDS related illness invariably result in periods of sickness and dependency. This result in disempowerment and lowered self-esteem, and decreased wellbeing amongst ill people.

Winnie et al (2007) used a prospective design in the study to test the attributional pathway from perceived control of responsibility to self blame and finally to self stigmatization and to examine the social and psychological sequelae of stigma amongst a sample of 119 people with HIV in Hong Kong, Structural equation modeling findings showed that the model had good fit to the data. Although the linkage between the attributions of control, responsibility and blame as confirmed the relationship of blame to self stigma was not significant. Self stigma was found to dampen social support and lead to psychological distress half a year later.

Stefanidis et al (2013) conducted a study on social anxiety and HIV/AIDS. They studied the links between HIV anxiety and other co morbid symptoms that manifest with it. They introduced the concept of stigma and how it can influence anxiety symptoms and then moves to describe how people living with HIV are perceived in their social environment. Also studied the role in affecting the behavior and psychological wellbeing of HIV patients. They looked at how much support within the family influences the feelings of anxiety and depression of HIV patients .They evaluated the intervention programs designed to reduce social anxiety and stigma. In absence of fully effective anti toxins remedies for HIV, factors such as safe sex practice, rising awareness and making people more knowledgeable about the disease have been assessed to better understand how effective they are in helping people better cope with their illness.

Hurley, Patrica, Ungavarski & peter (2008)conducted a study to explore the mental health needs of adults with HIV/AIDS on admission to homecare. A retrospective chart review of 244 adults discharged from a certified home health agency in New York City was completed. The majority of subjects were racial minorities and I/V drug users. HIV encephalopathy and toxoplasmosis were most frequently diagnosed Central Nervous System related AIDS indicator disease. Nurses most often noted symptoms of memory deficit was significantly related to inability to manage treatment medication non compliance and depression.

Comulada et al (2010) conducted a study to find the relationships between mental health symptoms (anxiety and depression) or a positive state of mind and behavior associated with HIV transmission (Substance use and risky sexual behavior) were explored in a longitudinal study of persons living with HIV who were participated in a transmission prevention trial. Bivariate longitudinal regressions were used to estimate the correlation between mental health symptoms and HIV related transmission acts for 3 time frames: at the baseline interview, over 25 months and from assessment to assessment. At baseline, mental health symptoms were associated with transmission acts. Elevated levels of mental health symptoms at baseline were associated with decreasing alcohol or marijuana use over 25 months. Over 25 months, an increasingly positive state of mind was associated with decreasing alcohol or marijuana use, an increasingly positive state of mind in the immediate intervention condition and increasing depressive symptoms in the lagged condition were related to increasing risky sexual behavior.

Mellins et al (2002) conducted a study on patterns of HIV status disclosure to perinatally HIV infected children and subsequently mental health outcomes. This study reports suggest that many HIV infected children, particularly these younger than 13 years, do not know they are HIV infected owing to parental concerns about the impact on their mental health. This study examined patterns of HIV status of disclosure to 77 perinatally HIV infected ethnic minority children (aged 3-13 years) and explore the association between knowledge of HIV status and emotional and behavioral outcomes. The majority of children in this study (70%) did not know their HIV status. On average children who knew their HIV status were older and tended to have lower CD4 %. Child knowledge of HIV status was not associated with gender, ethnicity, caregiver education, parent child relationship factors, type of placement (biological Vs. Adoptive) or other health status indicators. HIV status disclosure to infected children did not result in increased mental health problems.

McCains et al (1996) studied the influence of stress management training in HIV disease. They used a pre test post test design (with a 6 week wait list and 6 month comparison group) to compare the effectiveness of a 6 week stress management training program with standard outpatient of 45 men with HIV disease. The intervention consisted mostly of relaxation techniques, mental imagery and meditation .Outcomes included stress levels, coping patterns, quality of life psychological distress, illness related uncertainty and cd4 lymphocyte levels. At 6 weeks, intervention was associated with increases in the emotional wellbeing dimension of quality of life. After 6 months, the intervention groups had relative decline in HIV related intrusive thinking, indicating that stress management training may have buffered illness related psychological distress over time.

II. Literature related to Emotional and Behavioral Problems of PLHIV.

Banerjee [2007] conducted a study on behavioral disorders in Human immunodeficiency virus infected adolescents in the age group of 12-16 yrs in India. A prospective study was planned to examine the unique and combined influences of HIV and socio demographic characteristics on behavior of 441 children [140 infected and 301 matched controls] for which random sampling approach was used. Child behavior checklist was utilized in assessing behavior patterns in each child. Multivariate analysis comparing children and adolescents with HIV infection with their uninfected peers from similar background showed more subjective distress in HIV infected children. 80.7% HIV infected children reported behavioral problems by their primary caregivers as compared to 18.3% of healthy control group.

Naga seema (2008) conducted a study on depression among HIV patients. The effect of psychosocial intervention. Her study was conducted among HIV patients in King George Hospital in Visakhapatnam. The total sample includes 54 HIV/AIDS patients. The Beck depression Inventory was used for the study. The design used is pre and post test design. Before intervention the results indicated that 24% of HIV patients were extremely depressed, 26% were severely depressed. She had given counseling and psychotherapy to extremely depressed cases and found it was effective.

Mendoza, Hernandez, Castillo & Posner (2007) studied on Behavioral symptoms of children with HIV infection living in the Dominican Republic. The purpose is to describe behavioral problems encountered in a group of Dominican children living with Human Immunodeficiency Virus / Acquired Immunodeficiency Syndrome (HIV/AIDS) in the Dominican Republic. They were not receiving antiretroviral treatment. The participants were 43 children with HIV infection (2 to 8 years of age) who were attending an immunology clinic in the largest pediatric hospital in the Dominican Republic. All of the participants were vertically infected with the HIV virus (mother-to-child transmission) and had a very low socio-economic status. The children's caregivers were administered the Child Behavioral Checklist (CBCL) by trained psychologists to determine the caregivers 'perception of the children's behavioral problems. Behavioral findings were examined according to the CBCL age format: younger children (under 5 years of age) and older children (over 6 years of age) Descriptive statistics revealed a high proportion of the children, both younger (approximately 40%) and older (46%) scored in the borderline/clinical ranges for internalizing problems, including anxiety, withdrawn-depressed and somatic complaints. In addition, 46% of the older children were perceived as having externalizing problems (rule breaking and aggressive behavior).These findings suggest that a high incidence of behavioral and mood problems may be prevalent among Dominican children with HIV. The findings are discussed in terms of future research to examine other risk factors that might contribute to the high rate of maladaptive behaviors observed in the present report, including the contribution of socio-economic status, caregiver illness, caregiver education and parental loss.

Koopman et al (2007) studied on Psychological and Socio-medical Aspects of AIDS/HIV Relationships of perceived stress to coping, attachment and social support among HIV-positive persons. The purpose of this study was to examine the relationships of coping, attachment style and perceived social support to perceived stress within a sample of HIV-positive persons. Participants were 147 HIV-positive persons (80 men and 67 women). Multiple regression analysis was used to examine the relationships of the demographic variables, AIDS status, three coping styles, three attachment styles and perceived quality of general social support with total score on the Perceived Stress Scale (PSS). PSS score was significantly associated with less income, greater use of behavioral and emotional disengagement in coping with HIV/AIDS, and less secure and more anxious attachment styles. These results indicate that HIV-positive persons who experience the greatest stress in their daily lives are those with lower incomes, those who disengage behaviorally/ emotionally in coping with their illness, and those who approach their interpersonal relationships in a less secure or more anxious style.

Eley et al (2003) studied the genetic and environmental Effects on aggressive and rule-breaking behavior: Longitudinal Findings for Swedish twins in childhood and adolescence diagnostic systems such as DSM-IV typically combine aggressive behavior problems with rule-breaking behavior into a single diagnostic category of conduct disorder. However, empirically based studies have identified a syndrome of aggressive behavior (e.g., fighting, attacking people) and a separate syndrome of rule-breaking behavior (e.g., lying, stealing, truancy). Although aggressive behavior often correlates significantly with rule-breaking behavior, longitudinal studies have shown that the two syndromes follow different developmental paths. In addition, genetic factors have been found to influence aggressive behavior more than rule-breaking behavior. To illuminate relations between genetic factors and the developmental courses of the different kinds of conduct problems, Eley, Lichtenstein, and Moffitt (2003) performed genetic analyses of CBCL aggressive Behavior and delinquent Behavior (now called Rule-Breaking Behavior) syndrome scores for over 2,000 Swedish twins at ages 8-9 and 13-14. The genetic analyses indicated "that continuity in aggressive ASB (antisocial behavior) was largely influenced by genetic factors, whereas continuity in nonaggressive ASB resulted from both genetic and shared environmental factors", because longitudinal data are seldom available for making decisions in specific cases, Eley et al. concluded that "it may be necessary to rely on the phenotypic heterogeneity in young people's antisocial behaviors in order to define diagnostic subtypes" and that their study "provides discriminant validity for the etiology of these behaviorally defined subtypes".

Louthrenoo, Oberdordef & Sirisantana (2014) conducted a study on psychosocial functioning in adolescents with perinatal HIV infection receiving highly active ART. They conducted a cross sectional study of perinatally HIV infected adolescent, aged 11 to 18 years. Demographically matched controls were also enrolled. The adolescents completed the youth self report, while the care givers filled out the child behavior checklist to determine emotional and behavioral problems. The sample included 50 HIV infected adolescents and 56 controls. The internalizing problems scores from the YSR were significantly higher in the HIV infected group than those in the control group. The total competence scores, from both the self report and the care giver report in the HIV infected group, were significantly lower than those of the control group. Although HAART has prolonged the survival of HIV infected children, they remain at increased risk of psychosocial problems as well as impaired social functioning.

Adewuya et al (2008) studied the relationship between depression and quality of life in persons with HIV infection in Nigeria. Despite the fact that two-thirds of all the people with HIV live in sub-Saharan Africa, little is known about the emotional state and quality of life (QOL) of subjects with HIV in this region. The objective of this study was to evaluate the association between clinical depression and quality of life in a group of HIV sero-positive subjects in Nigeria. Methods: Subjects with HIV infection (n = 87) completed a questionnaire detailing socio demographic and HIV related variables. The subjects were assessed for the diagnosis of depression using the Mini International Neuropsychiatric Interview (MINI) and their subjective health related quality of life (QOL) was assessed using the short version of the WHO quality of life scale (WHOQOL-BREF). Results: There were 25 (28.7%) subjects with diagnosis of depression. Lower educational level correlated with poorer QOL in all the domains of WHOQOL-BREF except the "social relationship" domain. Also, poor social support correlated with poorer QOL scores on domains of "physical health" and "social relationship" and presence of medical problems was significantly associated with poorer scores on domains of "physical health" and "psychological health."

Diagnosis of depression was significantly correlated with poorer QOL in all domains except the "social relationship" domain. Conclusion: Poorer health related QOL in Nigerian subjects with HIV was associated with depression, lower educational and socioeconomic levels, and poor social support. Early identification and referral of patients with depression needs to be incorporated into intervention programs designed for HIV infected individuals in this region.

Schebusch & Vawda (2010) conducted a study with the aim to examine variables associated with recently diagnosed HIV-infection as a self-reported attempted suicide risk factor. Method: The study cohort consisted of 112 general hospital in-patients who attempted suicide because of HIV-infection. All the patients were subject to a comprehensive mental state examination and administration of a semi-structured questionnaire to obtain biographical, socio demographic and other relevant information. Pearson uncorrected Chi-square (X^2) or Fisher's Exact Probability Test were used to analyze data utilizing Epicalc 2000, version 1.02. Results: Most patients expressed a heterosexual preference. The average age was 34.9 years. Females predominated and in both genders depression and substance abuse (mainly alcohol-related) accounted for the most common psychiatric diagnoses. Less than half of the patients were married. A partner relational problem was a statistically significant variable. The most prevalent co-morbid stressors were poor social support, fear of disclosure/stigmatization and socio-economic pressures Cognitive deficits included problems with cognitive flexibility, concentration and memory Based on estimated national suicidal behavior prevalence rates, a descriptive HIV-related attempted suicide rate of 67.2 per 100 000 and an increased risk for attempted suicide of 13.33% to 18.87% were calculated. Conclusion: HIV-infection can be an underestimated suicide risk factor. Effective management and prevention programs should include as imperatives early diagnosis of HIV-related suicidal behavior, recognition of underlying psychopathology, neuro cognitive deficits, associated stressors, the dynamics of partner relationship problems, as well as cultural awareness and sensitivity. Potential neuro cognitive complications that can act as additional risk factors require further research.

Guzman, Rodrigo, Cantu, Bermudiz & Alvariz (2012) conducted a study to examine the psychosocial profile of individuals suffering HIV-AIDS in a Mexican city through assessing such variables as emotions, life events, social support, and coping, as well as their social representations. Participants were eight patients receiving medical treatment. In order to identify target variables measures of central tendency were taken and compared emotions all the time of initial diagnosis and at present. Results show that the negative emotional impact upon initial diagnosis included sadness, fear, and anxiety. These emotions diminished with time, whereas positive emotions, as optimism, confidence, and trust improved through successive diagnosis phases. Economical and employment aspects predominated as need for social support over affective and health features. The main coping styles were active, based on struggling, focusing, and carrying on concrete actions to solve problems. It is concluded that there seems to be a gradual social representation from the initial diagnosis to later ones, which lends support lo the assumption that not knowing the diagnosis may negatively influence behavior and attitudes. Expressly integrating social support to the treatment of HIV-AIDS patients is strongly recommended.

A study conducted by Marhefka et al (2009) in South Florida on emotional and behavioral problems and mental health service utilization of youth living with HIV acquired perinatally or later in life. Participants completed the age appropriate youth or adult self report symptom checklists (Achenbach System of empirically based assessment) and reported their psychiatric treatment history. Psychiatric diagnosis were abstracted from medical records. The 164 participating youth living with HIV were black (81%), females (52%), heterosexually identified (62%) and perinatally HIV infected (60%). Thirty one percentage reported levels of internalizing (emotional), externalizing (behavioral) or overall symptoms consistent with clinical psychopathology. In multivariate analysis, questioning one's sexual identity was associated with greater internalizing problems, whereas identifying as bisexual was associated with greater externalizing problems (P 0.05). Symptoms were not associated with HIV transmission group. Participants with >=1 composite score within the clinical range were more likely to received >=1 psychiatric service. However, 27% with clinically elevated scores has never received psychiatric care. Results suggest that among youth with HIV, those who identify as bisexual or questioning are at greater risk for emotional and behavioral problems.

Hockman et al (2009) studied on depressive symptoms and modeled emotional distress in persons living with HIV disease in non metropolitan areas at 13 U.S. States. Participants (329) were enrolled in a randomized clinical trial of a telephone delivered coping improvement group intervention and 60% reported moderate or severe levels of depressive symptomatology on the beck depression inventory. Structured equation modeling indicated that participants who experienced more severe HIV symptomatology received less social support and engaged in more avoidant coping also experienced more emotional stress. Greater HIV related stigma and rejection by family led to more emotional distress with social support and avoidant coping mediating almost entirely the effects of former two variables. The model accounted for 72% of variance in emotional distress in non metropolitan persons living with HIV disease.

Kelly et al (1993) conducted a study on 142 persons with HIV infection were administered the center for epidemiological studies depression scale, high levels of depression were predicted by lower perceived social support, attributions that health was influenced more by chance, high risk sexual behavior practices and greater number of HIV illness symptoms and greater duration of time knowing of one's own positive sero status. Ongoing high risk sexual behavior practices were predicted by higher levels of recreational drug use and of depression. These finding highlight the need for improved mental health services for persons with HIV conditions.

Lowther et al (2014) reviewed the previous studies to assess the prevalence of depression and anxiety reported in the literature and the presence or absence of the experience of stigma among HIV positive people on ART. The prevalence data from retained studies were analyzed by study location and data quality, Five data bases were systematically searched from 1996 to Aug 2013 using a predefined search strategy .Sixty six original studies identified the prevalence of depression, anxiety and presence or absence of the experience of stigma. The mean point prevalence of depression was 33.6% (SD=19.47) with lower reported point prevalence in high income countries (25.81%) compared to low and middle income countries (41.36%). The one to four week period prevalence of depression was 39.79%. Similar in high income countries and low and middle income countries. The point prevalence of anxiety was 28.38%, with a higher prevalence in low and middle income countries (33.92%) compared with high income countries (21.53%) with a wide variability. The mean point prevalence of stigma was 53.97% and one year period prevalence 52.11%. Results showed that HIV positive patients on ART report a higher prevalence of depression and anxiety than the general population, which nursing assessment and practice should address over half of HIV positive people report experiencing stigma.

Olisah, Adekeye, Sheikh & Yusuf (2009) studied to determine the rate of alcohol related problems in patients with HIV/AIDS and its association with high risk behavior. A consecutive sample of 120 patients with HIV/AIDS attending the medical outpatient department in Ahmadu Bello University teaching hospital, Zaria was assessed. All participants were screened for alcohol related problems using the alcohol use disorder identification test. High risk sexual behavior was assessed using the HIV risk taking behavior questionnaire. Alcohol related problems were found in 28.3% of participants (10% have hazardous use, 3.3% had harmful use and 15% had alcohol dependence). There was a significant association between alcohol related problems and risky sexual behavior. Alcohol related problems are fairly common in people already infected with HIV/AIDS are associated with high risk sexual behavior. Thus screening and treatment should be part of an effective HIV intervention group.

Carter et al (2003) conducted a study to clarify the relationship between subjective cognitive complaints and neuropsychological functioning in 160 adults with HIV infection. Participants completed questionnaires assessing cognitive complaints, symptoms of depressions and HIV related medical symptoms. Neuropsychological tests included measures of attention, verbal fluency, psychomotor skills, learning, memory and executive skills. Structural equation modeling was used to test models of the relationships among cognitive complaints, mood and medical symptoms with neuropsychological functioning. The model indicated that although depressed mood ($B=0.32$, $P<0.01$) and medical symptoms ($B=0.31$, $P<0.01$) influenced cognitive complaints. Cognitive complaints were independently associated with poorer neuropsychological performance ($B=0.39$, $P<0.1$). Mood and medical symptoms were significantly correlated but were not significantly associated with neuropsychological skills.

Cote, Jose, Pepler & Carolyn (2002) examined the effects of cognitive therapy nursing intervention on improving the quality of life for HIV positive males. 59 hospitalized HIV positive male patients (mean age=40 years) received either of two short nursing interventions focusing on either cognitive coping skills or the expression of emotions. Subjects completed questionnaires prior to and following the interventions concerning mood, distress and anxiety. Results show that both interventions lessened negative effects immediately following the initial daily session, however neither interventions produced any positive effects. Subjects receiving both interventions reported decrease in distress Cognitive coping subjects reported decreases in intrusive ideation and decrease in anxiety from immediately before to immediately after each session. Conversely emotional expression subjects reported increases in anxiety. It is concluded that the cognitive coping skills nursing interventions is effective in helping to regulate the emotional response of HIV positive individuals to advanced disease.

Kaplon, Marks, Garcy, Mertens & Steven (1997) conducted a study to find the distress and coping among women with HIV infection. In a multiethnic sample of 53 women with HIV / AIDS nearly 40% reported clinically significant levels of depressive symptomatology and anxiety compared to non patient form, distress levels were higher among the Latin, African American and white SS who made up with HIV sample, prayer and rediscovery of self were their most frequent coping responses suggesting that clinicians working with HIV/AIDS not overlook the importance of spiritual faith and practices in adopting to HIV infection.

Joan et al (2003) studied on Substance use and mental health correlates of non adherence to antiretroviral medications in a sample of patients with human immunodeficiency virus infection. Data were drawn from a survey of a national probability sample of 2267 (representing 181,557) adults enrolled in the HIV Cost and Services Utilization Study. This study focused on 1910 patients who reported their antiretroviral medication adherence during the past week. Patients with depression (odds ratio [OR] = 1.7; 95% confidence interval [CI]: 1.3 to 2.3), generalized anxiety disorder (OR=2.4; 95% CI: 1.2 to 5.0), or panic disorder (OR = 2.0; 95% CI: 1.4 to 3.0) were more likely to be non adherent than those without a psychiatric disorder. Non adherence was also associated with use of cocaine (OR = 2.2; 95% CI: 1.2 to 3.8), marijuana (OR = 1.7; 95% CI: 1.2 to 2.3), amphetamines (OR = 2.3; 95% CI: 1.2 to 4.2), or sedatives (OR = 1.6; 95% CI: 1.0 to 2.4) in the previous month. Compared with patients who did not drink, those who were moderate (OR = 1.6; 95% CI: 1.3 to 2.0), heavy (OR = 1.7; 95% CI: 1.3 to 2.3), or frequent heavy (OR = 2.7; 95% CI: 1.7 to 4.5) drinkers were more likely to be non adherent. These associations could not be explained by demographic, clinical, and treatment factors. These findings suggest the need for screening and treatment for mental health and substance use problems among HIV positive patients to improve adherence to antiretroviral medications.

Eric, Burnam, Douglas & John (2001) conducted a study on psychiatric disorders and drug use among human immune deficiency virus infected adults in The United states. They enrolled a nationally representative probability sample of 2864 adults receiving care for HIV in the United States in 1996. Participants were administered a brief structured psychiatric instrument that screened for psychiatric disorders (major depression, dysthymia, generalized anxiety disorders, and panic attacks) and drug use during the previous 12 months. Socio demographic and clinical factors associated with screening positive for any psychiatric disorder and drug dependence were examined in multivariate logistic regression analyses. Nearly half of the sample screened positive for a psychiatric disorder, nearly 40% reported using an illicit drug other than marijuana, and more than 12% screened positive for drug dependence during the previous 12 months. Factors independently associated with screening positive for a psychiatric disorder included number of HIV-related symptoms, illicit drug use, drug dependence, heavy alcohol use, and being unemployed or disabled. Factors independently associated with screening positive for drug dependence included having many HIV-related symptoms, being younger, being heterosexual, having frequent heavy alcohol use, and screening positive for a psychiatric disorder. Many people infected with HIV may also have psychiatric and/or drug dependence disorders. Clinicians may need to actively identify those at risk and work with policymakers to ensure the availability of appropriate care for these treatable disorders.

Myers, Hector, Durvasula & Ramani (1999) assessed the prevalence of major psychiatric disorders in African men and women affected by HIV / AIDS and estimated the relative contribution of demographics, substance use, burden of social strain and low social support in predicting psychiatric risk. Data are from analysis conducted on a community sample of 234 African American men who have sex with men (75 HIV negative and 159 HIV positive) and a sample of 135 African American women (35 HIV negative and 100 HIV positive). Both samples were relative young and low social support. Results indicated a high prevalence of psychiatric disorders in both samples with anxiety spectrum disorders (38%) and mood disorders (23%) especially depression (20%) prevalent among the men and depression (20%) and significant symptoms of post traumatic stress disorder (50%) prevalent among women. Results testing the contributions of risk factors were consistent with expectations with role strains. Low income, current drug use and low social support making contributions among men. Among women psychiatric disorders were associated with low social support and low income with weak contributions from low education and marijuana use.

Bogart et al (2011) explored in their study on PLHIV exhibit more severe mental health symptoms including depression and post traumatic stress disorder symptoms than do members of general public. They examined whether perceived discrimination which has been associated with poor mental health. A total of 180 black MSM (Men having sex with men) living with HIV completed audio computer assisted self interviews that included measures of mental health symptoms and scales assessing perceived discrimination due to HIV sero status, due to race / ethnicity and sexual orientation. In bivariate tests, all three perceived discrimination scales were significantly associated with greater symptoms of depression and post traumatic stress disorder. The multivariate model of depression yielded 3 way interactive among 3 discriminating types indicating that perceived racial discrimination was negatively associated with depression symptoms when considered in isolation from other forms of discrimination, but positively associated when all 3 other types of discrimination were present in multivariate tests, only perceived HIV related discrimination was associated with post traumatic stress disorder symptoms. Finding suggests that some types of perceived discrimination contribute to poor mental health among PLHIV.

Brown, Rundel & McManis (1992) conducted a study to assess the prevalence of psychiatric disorders in early stages of HIV infection. A part of a military HIV screening program, 442 men was assessed for the presence of DSM-III–R defined psychiatric disorders and symptoms of anxiety and depression after notification of HIV seroconverion. Of them, 84.4% were in the earliest, asymptomatic stages of diseases at the time of interview 96% did not have AIDS). The structural clinical interview for DSM-III R and structural interview guide for the Hamilton anxiety and depression scales were used. HIV sero positive men were more likely than age matched men in the community to have current diagnosis of major depression (ages 18-44) and anxiety disorder (ages 25-44). Higher lifetime rates of major depression and alcohol use disorder and high current prevalence of sexual dysfunction

(21.7%) were noted. Study conclude that men who become HIV sero positive have high rates of mood and substance use disorders prior to knowledge of sero conversion, and that early in the course of HIV infection men are at risk for developing major depression, anxiety disorder and disorders of sexual desire.

Benjamin Atwine, Elizabet, Cartor, Grace & Francses Bajunirwe (2005) conducted a study on psychological distress among AIDS orphans in rural Uganda. The objective was to investigate the psychosocial consequences of AIDS orphan hood in a rural district in Uganda and to identify potential areas for future interventions. The study population consists of 123 children aged 11-15 years whose parents were reported to have died with AIDs and 110 children of similar age and gender living in intact households in the same neighborhood. Symptoms of psychological distress were assessed using the Beck youth inventories of emotional and social impairment. A multivariate analysis showed that orphans had greater risk for high levels of anxiety, depression and anger. Further orphans had significantly higher scores than non orphans on individual items in the Beck youth depression inventory that are regarded as particularly sensitive to the possible presence of a depressive disorder that is vegetative symptoms, feelings of hopelessness and suicidal ideation. High levels of psychological distress found in AIDS orphans suggest that material support alone is not sufficient for these children.

Raymond et al (2010) conducted a study on depression care and prevalence in HIV positive individuals in Florida. They found it is a common psychiatric disturbances in individuals with HIV infection. Associated with sad mood, disturbed sleep and appetite, and low energy, the symptoms of depression may be difficult to distinguish from those of the infection itself. Depression has enormous clinical significance in this group of patients not only for the misery and poor quality of life it causes but also for its negative effect on patients sexual risk behavior, substances use and medication adherences. Effective treatments for depression in this group include antidepressant dedication, individual and group psychotherapy and social support interventions.

Judith, Robkin, Glenn Wagner & Richard Rabkin (1997) investigated the prevalence and treatment of depressive disorders in HIV illness. Studies of current and past psychotherapy in community samples of gay men have revealed surprisingly high rates of lifetime depressive disorders, despite low rates of current depression. Investigators from 2 different centers in a New York City reported life time prevalence of depression in the range of 30% to 35%, compared with the total male population life time rate of about 13%. Given the current nature of the disorder, the findings suggest a greater vulnerability to future episodes of depression in these groups. It has been suggested, that HIV itself causes mood changes and that AIDS related dementia includes depression.

Demi et al (1996) found the associations of suicidality with socio-demographic characteristics, number of HIV related symptoms, perceived stigma, depressive mood, emotional distress and family cohesion were investigated in a sample of women with HIV infection of 214 women, 56% reported neither suicidal thoughts nor attempts since learning they were HIV infected, 31% reported thoughts but no attempts, and 14% reported both thoughts and attempts. Women who reported suicide thoughts reported more HIV related symptoms, more perceived stigma, greater depressive mood, more emotional distress and less family cohesion a did women who reported no suicidal thoughts, women who reported both thought and attempts did not differ from women who reported only thought on these variables. Family cohesion moderated the effects of symptoms on thoughts. Those who reported suicidal thoughts reported more HIV related symptoms only when family cohesion was relatively low.

Hays, Robert, Turner, Heather, Coats & Thomas (1992) examined the impact of human immunodeficiency virus related conditions on depression among 508 gay men participating in the San Francisco men's health study, a population based prospective study of single man (aged 25-54 years). The number of HIV related symptoms experienced significantly predicted depression cross-sectionally and 1 year later. Satisfaction with each of 3 types of social support they received was less likely to show increased depression 1 year later. Degree of satisfaction with informational support appeared especially critical in suffering the stress associated with experiencing HIV symptoms. These findings offer valuable insight in understanding the psychological needs of gay man confronting the acquired immune deficiency syndrome crisis and have important practical implications for designing mental health services to meet these needs.

Liang, Li-Jung, Ying, Ji & Guoping (2009) made an effort to go beyond traditional analytical models to explore the complex and interactive nature of family processes. A total of 79 families affected by HIV in China participated in study 79 living with HIV and 79 sero negative family members with in-person interviews conducted in 2009. The higher levels of depression symptoms and social support and family relations were observed for both PLHIV and their family members. Results from active partner interdependence models indicate that the depressive symptoms of PLHIV and their family members were positively correlated where either family relations measure or the social support measure was included in the model. Results highlight the link between family experience and individual wellbeing with implications for designing and implementing intervention for families impacted by HIV.

Endeshaw, Meheret, Juddwalson & Rawlins (2014) performed a mixed methods cross sectional study to examine the relationship between stigma of HIV/AIDS and depressive symptoms in Gondaz, Ethiopia. They interviewed patients who presented for routine care at Gondar university hospital during the study period examining depressive symptoms and HIV/AIDS related stigma using standardized measures. Multiple regression was used to assess the relationship between depressive symptoms , stigma and gender .of 55 patients included in this analysis ,63.6 were female and most participants had limited formal education(69%,less than 12th grade education).The majority repeated experiencing both stigma(78%) and depressive symptoms (60%) ranging in severity from mild to moderately severe. Higher levels of HIV related stigma were significantly associated with higher levels of depressive symptoms. Although gender was associated with stigma, it was not associated with depressive symptoms. Results suggest the importance of psychosocial issues in the lives of people with HIV in Ethiopia.

Udell, Wadiya, Donenberg, Geri, Emerson & Erin (2011) studied the impact of mental health problems and religiosity on African American girls' high risk. They investigated the relationship between religiosity, mental problems and to sexual risk behaviors that are condom use and number of partners. Participants were 80 sexually active African American girls in psychiatric care and their care givers. Results indicated differentiated relationships, depending on parent versus youth report. Mother's religiosity was positively related to girls condom use and not to girls' number of partners. Controlling for other predictors in the models, mother's religiosity explained as much as 15% of the variance in girls condom use. Whereas parent and adolescent reports of girl's depression / anxiety and rule breaking were positively associated with number of partners, reports of aggression were associated with having fewer partners. Neither parent nor youth reports of girls mental health problems were associated with condom use. Controlling for other predictors in the models, girls mental health problems accounted for as much as 31% of the variance in number of partners. Findings underscore the importance of adopting an ecological framework to understand both the risk and promote factors for sexual risk taking among troubled girls.

Ickovics et al (1996) examined a long term effects of HIV counseling and testing for women in longitudinal prospective study. Women when perceived HIV C&T community health clinics (n=106) and a comparison group of never tested women (n=54) were interviewed 5 times over 18 months. There was no change in risk behaviors as a consequence of testing: tested and untested women engaged in a high risk sexual behavior at baseline and 18 months later. Tested women reported more anxiety, depression and intrusive thoughts about AIDS than did untested women. Although tested women were more concerned about AIDS, their potential risk factors over the study period generally were equivalent to those for untested women. HIV counseling and testing should be considered one aspect of a broader program of HIV prevention. Identification of alternative interventions must be a public health priority.

Nir et al (2014) studied on mapping white matter integrity in elderly people with HIV. People with HIV are living longer as Combination Anti Retroviral Therapy (CART) becomes more widely available. However even when plasma viral load is reduced to untraceable levels, chronic HIV infection is associated with neurological deficits and brain atrophy beyond that of normal ageing. HIV is often markedly by cortical and sub cortical atrophy, but the integrity of brains white matter pathways have been progressively declined. Few studies focus on older cohorts where normal ageing may be compounded with HIV infection to influence deficit patterns. In the relatively large diffusion tensor imaging study, they investigated abnormalities in wm fiber integrity in 56 HIV+ adults with access to CART (mean age: 63.9 (+) (-) 3.7 years), compared to 31 matched healthy control (65.4 -+2.2years). Compared to healthy controls HIV patients showed pervasive FA decreases and diffusivity increased through white matter. In both patients and controls, fiber integrity measures were associated with NP summary scores. The greatest differences were detected in the corpus callosum and in the projection fibers of the corona radiata.

III. Literature related to association between psychological health and ART therapy

Greeff et al (2014) studied to identify the relationship between a quality of life and anti retroviral treatment. Data were collected through a qualitative exploration of the experiences of QOL of people living with HIV or AIDS in Africa since being on ART, as well as quantitative measurement obtained from the HIV/AIDS targeted quality of life instrument .This study was conducted in three African countries. Purposive Voluntary sampling was used to recruit PLHIV through mediators working in community based HIV/AIDS organization and health clinics. Interviews were analyzed through open coding and the quantitative data through descriptive statistics and the Cronbach's alpha coefficient. A much wider range and richness of experiences were expressed by QOL instrument.

Gonalez et al (2004) conducted a study to find the relationship between social support, depression, positive state of mind and medication adherence among HIV positive men who have sex with men (n=61) and women (n=29) on highly active antiretroviral therapy. Depression and positive state of mind were evaluated as potential mediators of the relationship between support and adherence. Cross sectional data showed that greater social support and positive state of mind related to better adherence whereas higher depression scores related to non adherence. Positive state of mind partially mediated the relationship between social support and adherence. Positive state of mind may be an important mechanism through which social support is related to better medication adherence in this population.

Tomita, Andrew, Garrett & Nigel (2014) studied the impact of antiretroviral therapy and health related quality of life among South African women. Concerns are often raised regarding potentially adverse effects of antiretroviral therapy on health related quality of life, but there is limited longitudinal data to prove this, building on prior investigation they examined ART on HRQOL among HIV infected South African women with extensive follow up in the CAPRISA 002 acute infection cohort study. Overall HRQOL and five sub domains physical wellbeing. Emotional wellbeing functional and global wellbeing, social wellbeing and cognitive functioning were assessed using the functional assessment of HIV infection instrument. Their analysis comparing FAHI scores between pre ART and ART phases using paired Wilcoxon signed rank tests and adjusted mixed effects regression models revealed improvements on ART in overall HRQOL and in PWB, EWB and SWB, but not in FGWB and CF. No long term adverse impact on ART and HRQOL was detected.

Rosen & Sydney (2014) evaluated the effect of ART on south African HIV patients economic wellbeing as indicated by symptoms, normal activities, employment and external support, during the first 5 years of treatment. Prospective cohort of study of 879 adult patients at public or non government clinics enrolled before ART initiation or on ART less than 6 months and followed for 5.5 years or less. Patients were interviewed during routine clinic visits. Outcomes were estimated using population averaged logistic regression and reported as proportions of cohort experiencing outcomes by duration on ART. For patients remaining in care, outcomes improved continuously and substantially, with all differences between baseline and 5 years statistically significant ($p<0.05$) and continued significant improvement between year 3 and year 5. The probability of reporting pain last week fell from 69% during the 3 months before starting ART to 17% after 5 years on ART and fatigue from 62 to 7%.The probability of not being able to perform normal activities in the previous week fell from 47 to 5 % and of being employed increased from 32 to 44%;difficulty with job performance among the employed fell from 56 to 6%. As health improved, the probability of relying on a care taker declined from 81 to less than 1% and receipt of disability grant, which initially increased, fell slightly overtime on ART. Results from one of the longest prospective cohorts tracking economic outcomes of HIV treatments in Africa suggest continuous improvement during the first 5 years of treatment, confirming the sustained economic benefits of providing large scale treatment.

Kumar, Virender, Encinosa & William (2009) examined whether the relationship between HAART medication adherence and antidepressant treatment varied with HAART regimen complexity. The analysis included 1,192 respondents to the HIV Cost and Service Utilization Study (HCSUS) who were taking HAART. Self-reported past-week HAART adherence, antidepressant use, current mental health status, and an aggregate measure of regimen complexity were used in the analysis. Regression models with interactions between antidepressant treatment of mental health problems, poor emotional well-being or depressive symptoms, and medication complexity were estimated to assess differential associations with adherence. Irrespective of antidepressant treatment, poor mental health status was negatively associated with HAART medication adherence. However, only untreated higher depressive symptoms were strongly associated with mal adherence to HAART medication (OR = 0.72 at $P < 0.05$). Medication complexity was strongly associated with mal adherence to HAART medication (OR = 0.96 with $P < 0.05$) in the model including emotional well-being; and weakly associated with mal adherence (OR = 0.97 and $P < 0.07$) in the model including depressive and/or anxiety symptoms. However, as HAART medication complexity increased, adherence was higher among individuals with poorer mental health but using antidepressants compared to those with better mental health (OR = 1.09 with $P < 0.05$ in the model including emotional well-being; OR = 1.09 and $P < 0.05$ in the model including depressive and/or anxiety symptoms), and compared to those with poorer mental health but not using antidepressants (OR = 1.09, $P = 0.08$ in the model including emotional well-being, and OR = 1.12, $P < 0.05$ in model including depressive and/or anxiety symptoms). In conclusion, while individuals with poorer mental health generally have poor HAART adherence, their adherence improved with the use of antidepressants as the HAART complexity increased.

Kathryn whether, Susan Reif, Rachel Whethen & Murphy (2008) explored in their study that PLHIV often have complicated histories, including negative experience such as traumatic events, mental illness and stigma. As the medical community in the United States adapts to managing HIV as a chronic disease, understanding factors such as these negative experience that may be associated with poorer adherence to treatment regimens, greater HIV risk behavior, and lower patient quality of life becomes critical to HIV care and prevention. In less wealthy nations, these issues are also critical for addressing quality of life as well as medication adherence in the areas where anti retroviral therapies are being made available. They found the traumatic events, mental illness, including depression, anxiety and post traumatic stress disorder prevalent among individuals with HIV disease. Traumatic events, mental illness, distrust and stigma have also been linked with poorer adherence to medication regimens and HIV risk behavior.

Linnemayr et al (2013) conducted a study to evaluate the impact of antiretroviral treatment on employment related outcomes using prospective ,longitudinal analysis .Starting in january2008,602 treatment naive clients in one rural clinic and one clinic in the capital kampala were interviewed about their medical history and psychological and socio economic adjustments at baseline and at months 6 and 12 .Half of the sample was eligible to receive ART, while the other half was also in HIV care ,but not yet eligible for ART ,therefore providing a comparison group that is similar to the treatment group in that its members are HIV positive and have made the decision to enroll in HIV care. We found improvement in general health ,reduction in the incidence of pain and health interfering with work, as well as improvements in work related self efficiency for both groups over time, but significantly more so for the group receiving ART treatment. At baseline less than half of the people in ART group worked, but after 6 months more than 3 quarters of them were working, surpassing the fraction of people working in the control group after 1 year, Another key finding of the study was the importance of mental health as a key mediator for employment related outcomes . These data indicate that ART clients experience greater improvements compared to pre ART clients and not only with regard to general health, but also restoring confidence in their ability to work as well as actual work status.

Wagner et al (2012) examined the effects of ART on depression and other mental health indicators, 602 patients (302 non ART, 300 ART) were followed for first 12 months of HIV care in Uganda, with assessments at entry into care and months 6 and 12.Mental health was assessed with measures of depression, hopelessness and internalized HIV stigma; physical health functioning was assessed as an explanatory variable. Results: 13% had clinical depression, 57% had elevated depressive symptoms and CD4 cell count was negatively correlated with measures of depression at baseline. Significant reductions in elevated depressive symptoms and hopelessness were observed in both the ART and non-ART groups, but the drop in depression was more among the ART patients in intention to treat multivariate analysis. When added to the regression models, change in physical health functioning predicted positive longitudinal change on measure of depression, hopelessness an internalized stigma, yet ART status remained a significant independent predictor of each. Most mental health benefits of ART were experienced in first 6 months of care. These findings demonstrate the mental health benefits of ART and HIV care, however in some people, mental health problems persist once physical health is stabilized in which mental health treatment may be needed.

Grimes & Erim (2008) conducted a study in an attempt to learn about how integrative treatments (conventional medical care with complimentary treatment) are experienced by patients as means to gain further understanding about specific factors that influence their sense of physical and emotional health and wellbeing. Recently mind, body and alternative treatment has been utilized more frequently in the treatment. Prevention and recovery processes of complex health pro M blems, including stroke, cancer, migraines, depression and HIV and found physiological and psychological symptoms are related remains limited. Through semi structured qualitative interviews with patients who are involved in treatment at an integrative treatment centre in San Diego, patients experiences of treatment their health status and its impact on their sense of identity as well as its impact on their methods of coping were explored. Challenges faced in the treatment process were also explored in the interviews. Additional quantitative data, assessing participants' quality of life scale was also studied. The results of this study indicate that participants want information and resources to help them cope with their illness. Many participants reported that pursuing integrative treatment offered them resources to become proactive in the treatment process, contributed to a sense of empowerment and leads to feeling valued and understood. Results from the WHO quality of life scale demonstrated that participants who endorsed a stronger sense of self, feeling empowered and proactive in qualitative interviews scored higher on the WHO QOL scale.

Vassallo, Matko, Durant & Jacques (2014) studied on "can high central nervous system penetrating anti retroviral regimens protect against the onset of HIV associated neuro cognitive disorders?" Objective was to assess charges overtime in neuropsychological test results (NPR) and risk factors among a regularly followed HIV infected–patients population. Prospective cohort of HIV infected patient randomly selected to undergo neuropsychological follow up. Patients were divided into 5 groups: normal tests, neuropsychological deficit, asymptomatic neuro cognitive disorders ,and HIV associated dementia, Demographic and background parameters including CSF drug concentration penetration effectiveness (CPE) score 2010 were recorded changes in NPR and associated risk factors were analyzed. 256 patients underwent neuropsychological tests and 96 accepted follow approximately 2 years later. The groups were comparable upon neuro psychological retesting, six patients improved, 31 worsened and 59 were stable. The proportion of patients with HIV associated neuro cognitive disorders rose from 26 to 45%; with ANIS and MNDS still mostly represented. Most patients initially diagnosed with HANDS remained stable, 5 of 25 showed clinical improvement and 3 deteriorated. Of 33 patients with normal tests, 4 deteriorated where as 24 of 38 with initial neuropsychological deficit had poorer NPR and contributed most of the new HAND cases. Patients with clinical deterioration had a lower CPE score both at inclusion (6.9 vs. 8.1; p=0.005) and at the end of follow up (7.2 vs. 7.8;p=0.08) than those with improved or stable performance. This was confirmed by multivariate analysis. Patients with higher cope scores upon inclusion and at the end of follow up were at lower risk of clinical worsening, suggesting that CART with better CSF penetration could protect against cognitive deterioration.

Weaver Kathryn et al (2003) tested a structured model that incorporated age, time since diagnosis, social support, coping and negative mood as predictors of medication adherence on HIV viral load on 188 men and 134 women on highly active antiretroviral therapy. The authors used psychosocial latent factors formed from baseline measure to predict latent factors of adherence as assessed by electronic motoring and self report and viral load defined by indicators assessed over a 15 month period. Results from the model indicate that greater negative mood and lower social support are related to greater use of avoidance oriented coping strategies. Use of these coping strategies by patients on HAART is related to poorer medication adherence and subsequently higher viral load. This model advances researchers understanding of the contribution of psychosocial variables in predicting treatment adherence and disease progression in HIV positive men and women.

Johnson, Elliott, Neilands, Morin, Mallory & Chesney (2006) studied on social problem solving model of adherence to HIV medications. This study tested on explanatory model of HIV medication adherence using a social problem solving framework. Associations of SPS with adherence are hypothesized to be direct and / or indirect via psychological health HIV adults. HIV adults were interviewed using validated measures of SPS, psychological health and antiretroviral therapy medication adherence. Structured equation modeling techniques were used to test hypothesized relationships and to evaluate overall fit of the model to the data. SEM supported an indirect association of SPS on adherence via psychological health among the 545 HIV positive adults included in the analysis. Overall, the findings resulted in a model of adherence that offered very good fit to the data and correctly classified 97% of the cases as adherent Vs. non adherent. Results support the use of SPS as a conceptual framework for understanding adherence to ART. Findings offer rationale and directions for SPS intervention to enhance adherence by improving psychological health. Such approaches, if effective, have the potential to positively impact psychological wellbeing and adherence, thereby maximizing clinical benefit from treatment, which is linked to lower mortality from AIDS.

Kagee & Ashraf (2010) studied on psychological distress among persons living with HIV, hypertension and diabetes. They compared the mean scores on Hopkins symptom checklist, a measure of emotional distress of South African patients living with HIV/AIDS, hypertension and diabetes. The differences among four groups were non-significant, suggesting that living with HIV/AIDS may be no more emotionally distressing than other chronic illness. In this research note, it is speculated that these findings may be accounted for in terms of high prevalence of HIV in South Africa, the recent availability of antiretroviral treatment resulting in HIV being seen as chronic illness rather than a death sentence.

Arnal, Rafael, Maria, Beatriz & sandra (2012) studied on adherence to treatment in Spanish HIV patient. In an age where treatment with antiretroviral medication has allowed people infected with HIV to enjoy a better quality of life than ever before and have a life expectancy that until recently was considered inconceivable, adherence to treatment becomes especially relevant. Consequently, there is a need to understand in depth all factors associated with poor adherence to treatment in order to try to prevent problems or resolve them through effective intervention. The aim of this research was to assess the possible existence of a psychological profile associated with good or bad adherence to antiretroviral treatment. The participants were 66 Spanish patients with HIV in an approximately equal gender ratio in which 67.7 % were adherent to treatment. All participants underwent a battery of tests, which allowed assessing the difficulties that arise from taking medication and other psychological variables, such as perceived social support, self esteem, anxiety, depression, illness behavior, degree of interference of the disease, and personality dimensions. The results showed that a psychological profile for non adherent patients exists, which is characterized by a poorly perceived social support, high levels of anxiety and emotional disturbance, high psychological perception of the disease, less negation of daily problems and high degrees of extroversion and neuroticism.

Deanna et al (2008) explored the influence of HIV treatment optimism on the sexual behavior of people living with HIV/AIDS receiving highly active antiretroviral therapy at public health clinics in Brazil. They also explored the psychosocial dynamics of participants' sexual relationships in order to understand more broadly how these factors influence the sexual behavior or PLHA and how they shape HAART related beliefs. Twenty three semi-structured, in depth interviews were conducted with three groups: heterosexual women, heterosexual men and men who have sex with men living with HIV who reported being sexually active in the previous year and were currently receiving HAART. They found that the availability of HAART was conceptualized as a rationale for unsafe sex among a minority of study participants and that this was more common among men than among women. Participant from all three groups expressed a considerable amount of fear and anxiety regarding behaviors such as disclosure and condom use as a result of the unexplored conflict between implementing these behavior and continuing with their strategies for social validation within the context of their sexual relationships. They conclude that short term information, education and communication interventions surrounding treatment optimism, disclosure and condom use are appropriate and necessary.

Berger Greenstein, Cuevas, Brady, Trezza, Richardson & Kean (2004) investigated for major depression in patients with HIV / AIDS and substances abuse. This study conducted from 2001-2004 and examined rates and types of depressive symptoms in a cohort of patients receiving HIV care at two urban medical centers. These patients were participants in an intervention study examining adherence and mental health in persons triply diagnosed with psychiatric disorders, substance use disorders and HIV/AIDS. They sought to examine the relationship of depression to patient adherence to antiretroviral medication regimens. Results obtained from structural clinical interviews and self report questionnaires indicated that participants experienced high rates of depressive symptoms and that 72.9 % of participants met criteria for major depressive disorder. These depressive symptoms and major depressive disorder appear to be prevalent for this group, as well as the observation that these symptoms are amenable to treatment.

IV. The Literature related to music therapy

Cheryl & Ellen (2006) investigated the effects of chord arts/music therapy on physiological, psychological, social and quality of life among underserved minority HIV/AIDS patients and their caregivers in Philadelphia. Patients and their caregivers participated in a specially designed choir that met weekly over a 9 month period. Immediately following the choir, small group music therapy sessions were provided to participants. Participants in the program resulted in improved psychological wellbeing, reduced isolation, enhanced social support and increased hope.

Martwig (2006) conducted a study on music therapy in the context of palliative care in Tanzania. In light of increasing numbers of terminally patients in Tanzania owing to HIV/AIDs and cancer, limited access to opiods, and a growing interest in palliative care support. This study looks at the application of music in this context. A qualitative study was conducted by questionnaire of 17 professionals involved in home based palliative care in Tanzania. Findings include beliefs about the power of music, how music is being used to bring comfort to dying patients, and the most important aspects of helpful music to many Tanzania palliative care patients. Music can powerfully affect body, mind and spirit. It is vocal music, which is an accepted therapeutic music tool used to bring comfort to the palliative care patients and their family members.

Kenn (2002) conducted a study with a purpose to describe the frequency and correlates of complimentary and alternate medicine (CAM) therapies used by people with HIV/AIDS to manage illness and treatment related symptoms. Data were collected from a convenience sample (N=422) of PLHIV. Demographic variables (eg: education, age and gender) were compared for those who reported using at least one CAM therapy. Females and African Americans used CAM more frequently. More than one third of the participants used CAM, and there were a total of 246 critical incidents of non traditional treatment use. It is possible that these non allopathic interventions may positively affect health related quality of life in persons with HIV by ameliorating or reducing the side effects associated with the diseases and its treatments.

Michael & Silverman (2013) studied in Minneapolis the effects of music therapy on self and experienced stigma in patients on an acute psychiatric unit. The purpose of this study was to determine the effects of music therapy on self and experienced stigma in acute care psychiatric inpatients using a randomized design with wait list control. Participants (N=83) were randomly assigned by cluster to one of three single session group based conditions: music therapy, education, or wait list control. Participants in the music therapy and education conditions completed only post test while participants in the wait list control condition completed only pretests. The music therapy condition was a group songwriting intervention wherein participants composed lyrics for "the stigma blues". Results indicated significant differences in measures of discrimination, disclosure and total stigma between participants in the music therapy condition and participants in the wait list control condition. From the result of this randomized controlled investigation, music therapy may be an engaging and effective psychosocial technique to treat stigma.

Stark Jodi & canradi (2013) studied to investigate the effectiveness of a music therapy intervention on mood, psychosocial wellbeing, and quality of life of breast cancer survivors. Music therapy group intervention was 6 weekly 90 minute sessions, incorporating both active (music making, such as singing or instrument playing, music improvisation) and receptive (music listening, music assisted relaxation and forms of music with guided imagery) techniques. The qualitative data acquired through the music therapy questionnaire and session videotapes provided evidence of perceived treatment benefits noted by the women in the study, as well as the researcher. Major themes gleaned from these data through qualitative approaches suggested that the impact of cancer affected all aspects of quality of life and extended into survivorship. Participants used music listening to impact mood and express emotion during cancer treatment and recovery. Music therapy techniques increased self awareness, emotional expression and insight, leading to therapeutic growth and change. Participation in music therapy group provided social support and validation, enhanced coping skills and served as an outlet for emotional expression. These findings support the efficacy of a music therapy group intervention on mood, psychosocial wellbeing and quality of life for 3 breast cancer survivors who were participants in the study. Positive benefits were documented for all quality of life domains, including physical, psychological, social and spiritual wellbeing and were sustained at least over a short period of time.

Kirby, Rosemary & Olle (2012) stressed their helping pediatric patients cope with painful procedures is a critical element in providing comprehensive medical care. Although pharmacologic agents such as analgesics, sedatives and amnesiacs can be and often are used, they do not address all the elements of a painful experiences and can have undesirable side effects. Because of this, there has been a concerted effort to find non-pharmacologic therapies that can be used to reduce reliance on pharmacologic agents. The use of music to help children and adolescents cope with pain has been the subject of considerable research. They discussed the different ways in which music is used in clinical setting and highlight the unique aspects of music therapy, as contrasted with other music/sound based interventions, that make it a particularly powerful modality in the care of paediatric patient.

Yinger, olivia swedberg, Gooding & lori [2014] conducted a study on the role of music therapy and music medicine for children and adolescents. Music is a useful tool in treating children and adolescents with psychiatric disorders because of the engagement of multiple areas of the brain that occur during music listening. The strength of music therapy as a treatment of children and adolescents with mental health concerns include the fact that music therapy is known to benefit clients with a wide variety of diagnosis and can be effectively administered in group format, making it a safe, cost efficient treatment. Neuro imaging research indicates that listening to preferred music activates rewards circuitry in the brain and active musical participation engages more areas of the brain than passive listening. Music therapy is an established health care profession in which music is used within a therapeutic relationship with a music therapist to accomplish non music goals.

Callaghan et al (2013) conducted a study to examine music relevance, including pre loss music therapy, for 8 informal caregivers of people who died from cancer. The design was informed by constructivist grounded theory and included semi structural interviews. Bereaved caregivers were supported or occasionally challenged as their musical lives enabled a connection with the deceased. Music was often still used to improve mood and sometimes used to confront grief. Specific music, however, was sometimes avoided to minimize sadness. Continuing bonds theory's focus on connecting with the deceased through memory and imagery engagement may expand to encompass musical memories, reworking the meaning of familiar music and discovering new music related to the deceased. Pre loss music involvement, including music therapy, between dying patient and families can help in bereavement.

Gallagher (2002) studied to assess the potential for music therapy among patients with a range of chronic and or advanced illnesses. She focused on 200 patients battling several types of cancer, non cancerous tumors, pain disorders, sickle cell disease, AIDS and other life limiting diagnosis. Patients were between the ages of 24 and 87. About 60% of patients were women, and the researcher noted that almost 30% had some sort of musical background. Music therapy first involved the patient choosing the style of music her on she wanted to hear. The therapy averaged approximately 25 minutes. Physical and psychological tests were conducted both before music therapy and after an initial therapy session. The results indicated that in addition to marked benefits in patient anxiety, mood, pain and shortness of breath, more than 80 percent of patients said their mood had improved following music therapy.

Archer, Buxton & Sheffield (2014) examined the effectiveness of creative psychological interventions (CPIS) for adult cancer patients. In particular the findings of randomized controlled trials of art, drama, dance/movement and music therapies on psychological outcomes were examined methods the review yielded 10 original studies analyzing data from a total of 488 patients' data extraction and quality assessment were conducted by two independent reviewers. Results four of papers focused on the use of art therapy, three studies used music therapy, one paper utilized dance therapy. One study used dance therapy and remaining paper used creative art therapies which was a combination of different art based approach. Eight papers focused solidly on breast cancer patients and the remaining studies included mixed cancer stages. The studies reported improvements in anxiety and depression, QOL, coping, stress, anger and mood .However a few physical benefits of CPI were reported, there was no significant impact of CPI on physical aspects of quality of life, vigor activity or fatigue inertia or physical functioning. There is a evidence that CPIS benefit adult cancer patients with respect to anxiety and depression, quality of life, coping stress, anger and mood.

Ahmadi & Fereshteh (2013) conducted a qualitative study in Sweden on music as a method of coping with cancer. A qualitative study was conducted based on semi structural interviews with 17 cancer patients. Participants were chosen from a group of patients who has listened to or played music as a means of coping with their illness. This study showed the importance of considering the roles of those different kinds of music play in coping with cancer. The music of nature, healing music, religious music and cheerful music each have different benefits for patients. A patient's situation and his individual characteristics determine the kind of music that can be used as useful or harmful coping strategy. Therefore it is essential to investigate the types of individual characteristics that can make listening to different kinds of music as a helpful or harmful coping method.

Pothoulaki Maria et al (2012) studied the use of music in chronic illness in Newyork. Music interventions have been applied to many of these health conditions, addressing the needs of individuals with cancer, mental illness, renal failure, multiple sclerosis, HIV, traumatic brain injury, chronic pain, cardiac disease and those in rehabilitation suffering from diverse muscular and neurological diseases. Results suggest beneficial effects of music listening upon a range of physiological (e.g. blood pressure, heart rate, enzyme production, respiration) and psychological variables (anxiety, mood, relaxation, pain). Theoretical integration and synthesis is then explored with 3 mechanisms presented as possible explanations for the positive effects of music listening, 1) music communication as a form of social support 2) emotional engagement with music 3) increased levels of perceived control.

Lee, eun-jeong, Bhattacharya & Joydeep (2013) investigated the impact of monochord (mc) sounds a type of archaic sound used in music therapy, on the heart rate variability in patients with cancer undergoing chemotherapy. The HRV of patients were recorded during the first and last sessions of relaxation treatments. The time series of HRV was analyzed using methods based on time and frequency domains. During MC induced relaxation, low frequency power and ratio of low and high frequency were increased over sessions. However, changes in low and high frequency lay in different directions during relaxation caused by progressive muscle relaxation in each session. Different activities of the parasympathetic and sympathetic nervous system were shown during relaxation caused by either listening to mc sounds or exercising PMR.

Callaghan, Clare, Dun, Beth, Baren, Annette, Barry & Philippa (2013) found in their study the music is central in most children's lives. Understanding its relevance will advance efficacious pediatric supportive cancer care. Qualitative clinical data mining uncovered four music therapist, perspectives about music and music therapy's relevance for pediatric oncology patients up to 14 years of age. Inductive and comparative thematic analysis was performed on focus group transcripts and qualitative inter rater reliability integrated. Music can offer children a safe haven for internalizing a healthy self image alongside patient identity. Music therapy can calm, relieve distress, promotes supportive relationships, enables self care and inspire playful creativity, associated with normalcy and hope. Preferred music and music therapy should be available in pediatric oncology.

Cordobes (1997) study used the activity of group song writing as a CBT for HIV patients and used two control groups of group games playing. They also had a third group that did not go through any kind of CBT activity. The results showed that more emotions were expressed in song writing activity and there is evidence that song writing can be used for clinical that song writing can be used for clinical usefulness. The study also concludes that song writing activity could be analyzed to be useful in increasing social support. He suggested that music therapy can be used in various forms for HIV patients and benefits can be multiple as well.

Edward & Carole (1990) studied the characteristics of motivation for music and musical aptitude of undergraduate non music majors. Non music majors in a music appreciation course were administered an experimental college version of the musical aptitude profile and two measures of motivation for music. One motivation measure assessed five factors to which students attribute success or failure in music: effort, musical ability, background, classroom environment and affect for music. The other measure assessed the magnitude of student motivation for music in three areas: personal commitment, school music and music compared to other activities. Low reliability obtained on the musical aptitude profile. Musical scale necessitated its elimination from the analysis. No significant differences due to sex or academic level were obtained, nor was there a significant relationship between the motivation and aptitude variables. The motivating factors and magnitude-of- motivation measures were significantly related, although these two types of variables were found to measure difficult aspects of music motivation. The main reason cited for being musically motivated was affect for music.

Stephens, Braithwaite & Taylor (1998) conducted a study on finding effective and preventive techniques of HIV among young African (AA) population. Authors suggested that hip hop music is popular and can be used as an effective tool to achieve the goal of prevention as well s therapy for young people who are at a higher risk as AA population have a higher chance of getting HIV.

[V] The literature related to influence of demographic variables on mental health of PLHIV.

Mcgowan et al [2014] conducted a study on effect of age on symptom burden, mental health and quality of life amongst people with HIV in U.K. It included 3258 HIV diagnosed individuals [2248] MSM : 378 heterosexual men, 637 women [64% response rate]. Participants completed a questionnaire that included distress, depression & anxiety and health related quality of life problems. Of all participants , 87% were taking ART, 76% had viral ≤ 50 c/ ml and 19% had CD4 < 350 /mm3 , Mean age was 45 yrs [range 18-88] with 5% < 30, 23% 30-39, 43% 40-49, 22% 50-59 and 7% more ≤ 60 Yrs. The most prevalent distressing physical symptoms were lack of energy [26%], difficulty sleeping [24%], muscle ache/joint pain [21%] and pain [18%]. With older age, there was no clear trend in prevalence of physical symptom distress, but prevalence of depression and anxiety decreased while prevalence health related QOL problems increased. This pattern remained after adjustment for gender / sexuality and time diagnosed with HIV. The increase with age in over all prevalence health related QOL problems was due to increased problems for mobility, self care and performing usual activities domains, not an increase in depression/ anxiety. Longer time with diagnosis HIV was strongly associated with higher prevalence of all symptoms measures and Health related QOL problems. Physical and psychological symptoms are common among people living with HIV, but the burden of these symptoms are not highest among the older age group. While HLQOL tended to worsen with old age, physical symptoms distress did not and mental health improved. This may reflect greater resilience in older adults, or the potential for successful aging maintaining mental health despite age related health losses.

Kapadia et al (2016) studied on perceived social support and mental health states in a clinic based sample of old HIV positive adults. Despite a high prevalence of mental health problems in older, HIV adults [aged >50], few studies have examined the relationship between perceived social support and mental health burdens in this group. In a clinic based sample of 100 adults, 23% and 29% met criteria for one and more than one mental health burden, respectively. In multivariable binary logistic models, perceived receipt of positive and emotional social support as well as overall support was inversely associated with the presence of mental health burdens.

Liang, lee & farmers [2012] conducted a study on HIV status and gender in Thailand. They examined the relations among HIV states, gender and depressive symptoms among 152 married or cohabitating couples living with HIV in the northern and north eastern regions of Thailand. Depressive symptoms were assessed using a 15 item depressive symptom screening test that was developed and used previously used in Thailand. Among 152 couples, 59% were couples in which both members were living with HIV [seroconcordant], 28% had only female living with HIV [serodiscordant] and 13% had only male members with HIV [serodiscordant]. The prevalence of depressive symptoms between seroconcordant and serodiscordant groups was similar. However females living with HIV reported significantly higher levels of depressive symptoms, regardless of their partner's HIV status.

Raboud & Levis [2012] studied on influence of employment and job security on physical and mental health in adults living with HIV. They used baseline data for 1660 men and 270 women who participated in the Ontario HIV treatment network cohort study, an ongoing observational cohort study that collects clinical and socio behavioral data from people with HIV in the province of Ontario Canada. They performed multivariable regression analysis to determine the contribution of employment and job security to health related quality life after controlling for potential confounders. Employment men with secure jobs reported significantly higher mental health related quality life than those who were non employed, but insecure employment was not associated with higher mental health scores relative to non employment. Among women, being employed was significantly associated with both physical and mental health quality of life, but job security was not associated with additional health benefits. Participation in employment was associated with better quality of life for both men and women with HIV.

Summary of review literature

The literature reviewed clearly demonstrated that HIV infection is a chronic debilitating disease and antiretroviral therapy can improve the quality of life of PLHIV. Many research studies such as Heywood & Lyons, Charney et al, Gupta et al, Felicity Thomas, Hurley patrica et al, Claude Anne Mellins focused on mental wellbeing of PLHIV as HIV diagnosis, treatment, its side effects, stigma and opportunistic infections could affect the mental health of HIV infected. Good mental health is essential for successful treatment (ART) of PLHIV.

A number of empirical studies on emotional and behavioral problems of PLHIV indicated that they are at risk for developing these problems due to negative responses towards stress. Tanusree Banerjee, Nagaseema, Louthrenoo et al, Guzman et al, Marhefka, Raymonds, Judith, Li Li Liang studied upon emotional and behavioral problems. These emotional and behavioral problems can affect both physical and mental health of PLHIV.

Mcgown et al study related to effet of age on mental health, Kapadia et al focused on the relationship between social support and mental health, Lianj, Lee & Farmer studied the gender and HIV status, Roboud &Levis study on influence of employment on physical and mental health of HIV adults proved that age, gender, employment and social support have greater influence upon mental health of PLHIV.

Tomito Andrew et al, Rosen,Sydney et al, Arnal Rafael Ballester et al, Johnson Mellory et al, Wagner,glenn et al, Greef, Minrie et al studies focused on impact of ART on Psychological health and proved that ART could improve the quality of life. There are studies related to impact of music therapy on psychological health and quality of life such as Gallagher, Hartwig, R.,Kenn M. Kirsey, Pothoulaki Maria et al, Dr Cheryl Dileo and Ellen Tedaldi, Corbodes, Stephens focused and proved that music therapy could improve the psychological health of PLHIV.

One deficiency in the current literature base is that many studies featured predominantly on assessing mental health, emotional and behavioral problems and the impact of ART and music therapy on Psychological health of PLHIV, but the association of music appreciation with mental wellbeing, emotional and behavioral problems among PLHIV on Antiretroviral therapy have not been investigated. So the present study focused on it.

CHAPTER - III

METHODOLOGY

The methodology adopted for the present study is described in this chapter. It deals with the objectives of the study, research design, sample, measures, variables, hypothesis, data collection procedure and statistical technique used for analyzing the data.

Objectives:

1. To assess wellbeing, emotional and behavioral problems, appreciation of music among PLHIV on Antiretroviral therapy.
2. To find out the relationship between appreciation of music and emotional behavioral problems among PLHIV on Antiretroviral therapy.
3. To find out the relationship between appreciation of music and wellbeing among PLHIV on Antiretroviral therapy.
4. To find out the relationship between emotional behavioral problems and wellbeing among PLHIV on Antiretroviral therapy.
5. To find out the relationship of demographic variables with emotional and behavioral problems, wellbeing.

Method

Research Design:

Descriptive survey design was found appropriate as surveys are carried out for the purpose of providing an accurate portrayal of a group of subjects with specific characteristics. For this study, descriptive survey design was used to identify the emotional and behavioral problems, wellbeing and appreciation of music among PLHIV on Antiretroviral therapy and to find association between them.

Setting of the Study:

The setting selected for the present study was ART Centre at King George Hospital, Visakhapatnam and ART Centre at Osmania General Hospital, Hyderabad in united Andhra Pradesh which were found to be the biggest centres and enough sample could be selected.

Sample:

The sample for the present study was 287 PLHIV attending ART centres at King George Hospital, Visakhapatnam and Osmania General Hospital, Hyderabad. The age range of sample is from 18 years to 59 years with a mean 36.86. The sample characteristics were also explained in terms of age, gender, education, occupation, income, place of living, marital status, caste, religion, number of dependents, mode of transmission, living with partner, HIV disclosure status, duration of ART and family and social support.

Sampling Technique:

For this study the purposive sampling technique was used to select the sample. It is one of the Non Probability Sampling method, which is based on researcher's knowledge about the population and is used to hand pick the cases that meet the criteria set by the researcher.

Inclusion and exclusion criteria:

The study included adults of both male and female whose age falls between 18-59 yrs of people living with HIV undergoing Anti retroviral therapy. Those PLHIV on ART whose age is below 18yrs and above 59 yrs and not on Antiretroviral therapy were excluded from the study.

Measuring Tools:

1. **Wellbeing Scale**:

The WHO Bradley wellbeing questionnaire, developed by Clare Bradley [1994], was used to measure the psychological wellbeing of the sample. The wellbeing questionnaire is a 22 item, multidimensional measure that assesses depression (6 items), anxiety (6 items), energy (4 items) and positive wellbeing (6 items).total wellbeing can also be calculated by combining the subscales. The each item is scored on a 0 to 3 Likert scale where 0 indicates that the respondent felt that the item had applied to them "not at all' over the past few weeks while 3 indicates that it applied "all the time.". Scores of 1 and 2 indicates points between the extremes. Ratings for item on each subscale should be summed after reversing scores where necessary (Depression items 1,3,4 and 6;Anxiety items 11 and 12; Energy items 14 and 15).Subscales are scored that a higher score on each subscale indicates more of the mood described by the subscale scale i.e. depression, anxiety, energy and positive wellbeing. A general wellbeing total score can be obtained by summing the subscale scores after reversing the scores on the depression and anxiety subscales. The items are particularly sensitive to the more cognitive aspects than to the somatic aspects of the scale's dimensions. Originally, the scale was developed for use in a W.H.O. study evaluating new treatment for the management of diabetes. However, as Bradley stated the scale can be used with general population also as it does not include any item that is particular to diabetes. The researcher used the Telugu version of the Bradley wellbeing questionnaire to assess the wellbeing of PLHIV on antiretroviral therapy

Reliability:

Scale reliability alpha ranged from 0.70 to 0.88 with good internal consistency.

Validity:

The product movement correlation (validity coefficient) 0.57 was significant (p=0.001). This established concurrent validity.

2. Adult Self Report [ASR]:

The original English and developed Telugu version of adult self report [ASR] was used for the present study. The ASR is component of ASEBA that have been developed based on decades of research and practical experience since 1960's [Achenbach, 1966], ASEBA forms capture both the similarities and differences in how adults function under different interaction partners.

The adults self report consists of 126 problem item checklist developed by Achenbach and Rescorla [2003], that enable the people (18-59 years) to report the degree of severity in each problem by answering '0' if the item is not true of himself or herself, '1' if the item is somewhat or sometimes true, and '2' if it is very true often true within the past 6 months. ASR form can typically be completed in about 15 to 20 minutes. The ASR include eight empirically based syndrome scales anxious / depressed, withdrawn, somatic complaints, thought problems, attention problems, aggressive behavior, rule breaking behavior and intrusive behavior. The first five syndromes form a grouping designated as internalizing, while the last three syndromes form a grouping designated as externalized. The total problems score is computed by summing the 1 and 2 scores on all the problem items. High scores on the problem scales are clinically significant, because they indicate high levels of problems.

Reliability:

Reliability was generally very high, with all test- retest 'r' being significant at 0.01 and most being in the 0.80 and 0.90 with none <.71. Good internal consistency was found for most scales for mean alpha coefficient on the ASR and ABCL of 0.83 and 0.85 for empirically based problem scales, 0.78 and 0.79 for the DSM oriented scales

Validity:

Correlation of 0.81 for the total problems score and the externalizing scale and 0.70 for the internalizing scale demonstrating excellent validity.

3. Music appreciation scale:

Music appreciation scale was developed by Dr. Gowri Ram Mohan (2000). The English version of the tool was given to two university members who are proficient in both English and telugu languages. First, the tool was translated into Telugu and retranslated into English and finally the Telugu version of the tool was used by the researcher. It measures both the music interest and music aptitude. It consists of 22 items. Each item was given options scoring from 1 to 3.Among 22 questions, 11questions such as 1,2,3,4,6,8,9,11,12,14,17 measures music interest and remaining questions 5,7,10,13,15,16,18,19,20,21,22 measures music aptitude.

Reliability:

Odd and Even method yielded a correlation of 0.77. Split half method yielded a correlation of 0.73. Test and Retest method yielded a correlation of 0.74. concurrent validity was found high.

Variables:

Independent Variable: Music appreciation. Further Age, gender, education, occupation, income, place of living, marital status, caste, religion, number of dependents, mode of transmission, living with parents, HIV disclosure status, duration of ART, family and social support are the attributes also act as independent variables.

Dependent Variable: Wellbeing, emotional and behavior problems.

Hypothesis:

1. There will be a significant relationship between interest in music and emotional and behavioral problems of PLHIV on ART.
2. There will be a significant relationship between interest in music and wellbeing of PLHIV on ART.
3. There will be no significant relationship between music aptitude and emotional and behavioral problems of PLHIV on ART.
4. There will be no significant relationship between music aptitude and wellbeing of PLHIV on ART.
5. There will be no significant difference between demographic variables and emotional behavioral problems of PLHIV on ART.
6. There will be no significant difference between demographic variables and wellbeing of PLHIV on ART.

Ethical considerations:

1. Investigator obtained permission from NACO, Delhi before starting to conduct a study on PLHIV.
2. Investigator obtained the institutional Ethics committee approval prior to the study.
3. Informed consent was taken from PLHIV after explanation of the purpose of the study.
4. Investigator assured the participants regarding the confidentiality of the personal information shared during the data collection.

Data Collection Procedure

The researcher obtained permission from National AIDS Control Organization, New Delhi and Andhra Pradesh state AIDS control society, Hyderabad to conduct a study on PLHIV. The researcher also obtained Ethical Committee approval from Institutional Ethics Committee, King George Hospital, Vishakhapatnam. Prior to the study formal permission was taken from the medical superintendent of King George Hospital, Vishakhapatnam and medical superintendent of Osmania General Hospital, Hyderabad to collect data from PLHIV attending ART centres of both the hospitals. After obtaining consent from the PLHIV, researcher collected the data from PLHIV.

All the participants were assured of the confidentiality of their answers. The participants who could read and write were given all the three questionnaires at a time. No time limit was set for completing these three forms. From the participants who cannot read and write, the researcher collected the data through face to face interview. It took approximately 40 to 50 minutes for each participant to complete.

Analysis of Data:

Data analysis was carried out by using the SPSS 14.0 version. Descriptive statistics was used to describe the data and the inferential statistics was used to test the hypothesis. Pearson Correlation was used to explore the relationship between appreciation of music and emotional behavioral problems as well as music appreciation and wellbeing of PLHIV. The t-test and F-test (Analysis of Variance) were used to examine the influence of demographic variables on the emotional and behavioral problems and wellbeing.

CHAPTER - IV

RESULTS AND DISCUSSION

This chapter presents the results of the statistical analysis used to study the research objectives established in the previous chapters. First, demographic characteristics of the sample are presented. Second, the findings of the five research objectives are addressed.

Table-1

Sample Characteristics according to Gender, Age, Education and Occupation

Variable	Total Sample	Groups	Frequency	%
Gender	287	Male	176	61.3
		Female	111	38.7
Age	287	Below 30 Years	80	27.9
		30 to 40 Years	130	45.3
		Above 40 Years	77	26.8
Education	287	Illiterate	100	34.8
		Tenth and Below	144	50.2
		Inter and Graduation	43	15.0
Occupation	287	Self Employee	41	14.3
		Laborer	69	24.0
		Employee	59	20.6
		House Wife	54	18.8
		Business	21	7.3
		Farmer	22	7.7
		Grade-IV Employees	21	7.3

Table 1 shows that out of 287 participants, 61.3% (n=176) were males and 38.7% (n=111) were females. About their age, majority 45.3% (n=130) were in the age group of 30- 40 years. 27.9% (n=80) were in the age group of below 30 years and remaining 26.8% (n=77) were above 40 years. Regarding education it is observed that the highest 50.2% (n=144) studied 10th and below, 34.8% (n=100) were illiterates and only 15% (n=43) had completed intermediate and graduation. With regard to occupation, the findings shows that 29% (n=69) were laborers, 20.6% (n=59) were employees. Remaining were from business, farmers and grade IV employees contributing each around 7% respectively.

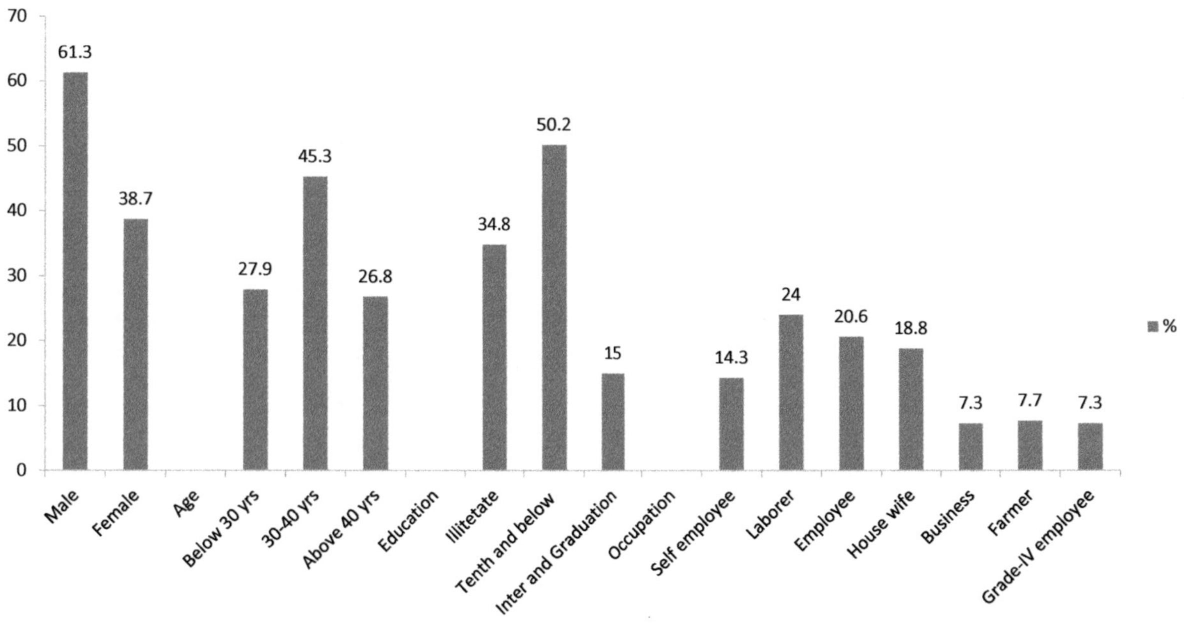

Figure 2: Sample characteristics according to Gender, Age, Education and Occupation

Table-2

Sample Characteristics according to Income, Place of Living, Marital Status, Caste and Religion

Variable	Total Sample	Groups	Frequency	%
Income	287	Below Rs 3000	96	33.4
		Rs3000- Rs6000	118	41.1
		Above Rs 6000	73	25.4
Place of Living	287	Urban	233	81.2
		Rural	54	18.8
Marital Status	287	Married	230	80.1
		Single	18	6.3
		Married but Single	39	13.6
Caste	287	SC&ST	44	15.3
		BC	173	60.3
		OC	70	24.4
Religion	287	Hindu	224	78.0
		Others	63	22.0

Table 2 shows that out of 287 participants, the majority 41.1% (n=118) were in the income of Rs.3000-6000, and 33.4% (n=96) were in the income of below Rs. 3000 and only 25.4% (n=73) were having above Rs 6000 per month. With regard to place of living majority 81.2% (n=233) were living in urban area and only 18.8% (n=54) were from rural area. About marital status the highest 80.1% (n=230) were married and only 6.3% (n=18) were unmarried. Regarding the caste, the findings shows that the highest 60.3% (n=173) belongs to BC, 24.4% (n=70) belongs to OC and only 15.3% (n=44) were from SC and ST. About religion, majority 78% (n=224) were Hindus and remaining 22% (n=63) were from other religions.

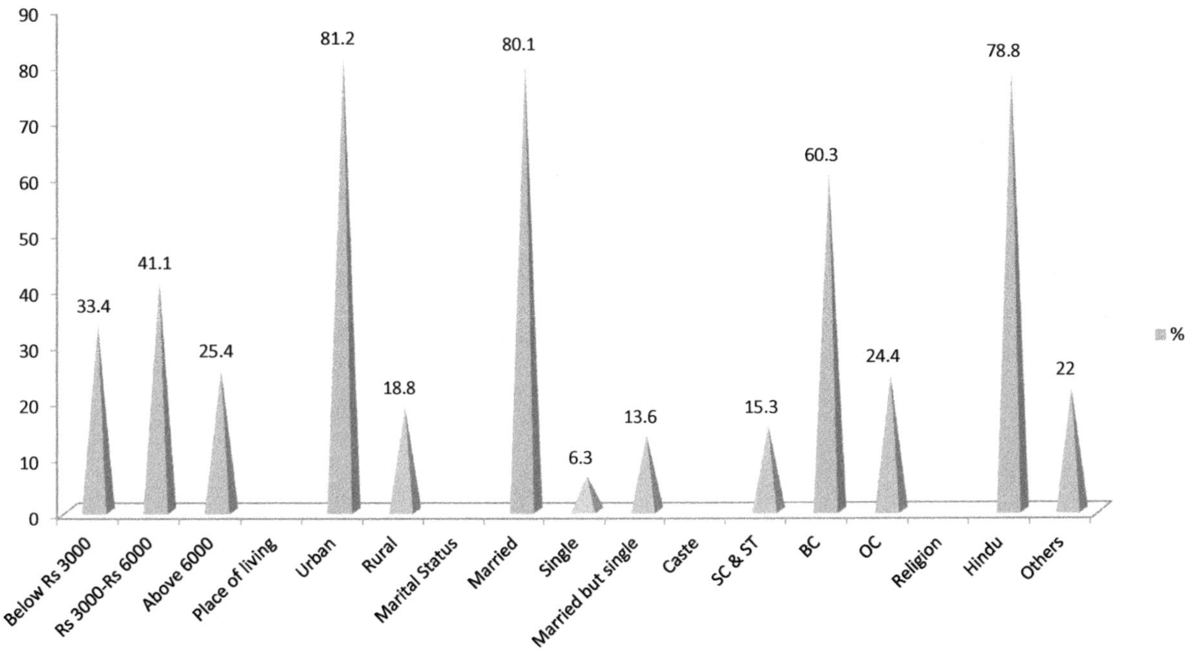

Figure 3: Sample characteristics according to Income, Place of living, marital status, Caste and Religion

Table-3

Sample Characteristics according to Number of Dependents, Mode of Transmission and Living with Partner

Variable	Total Sample	Groups	Frequency	%
No. of Dependents	287	No Dependents	15	5.2
		One or Two	125	43.6
		More than Two	147	51.2
Mode of Transmission	287	Sex	202	70.4
		Injections	2	0.7
		Blood Transmission	3	1.0
		By Birth	1	0.3
Living with Partner	287	Yes	203	70.7
		No	84	29.3

Table 3 shows that out of 287 participants, majority 51.2% (N=147) were having more than 2 dependants and 43.6% (n=125) were having 1 or 2 dependants and very less 5.2% (n=15) were having no dependants. Regarding mode of transmission, the highest 70.4% (n=202) infected through sex, 27.5% (n=79) had unknown status and remaining 2% were infected through blood transmission and by birth. With regard to living with partner the highest 70.7% (203) were living with partner and remaining 29.3% (n=84) were not living with partner.

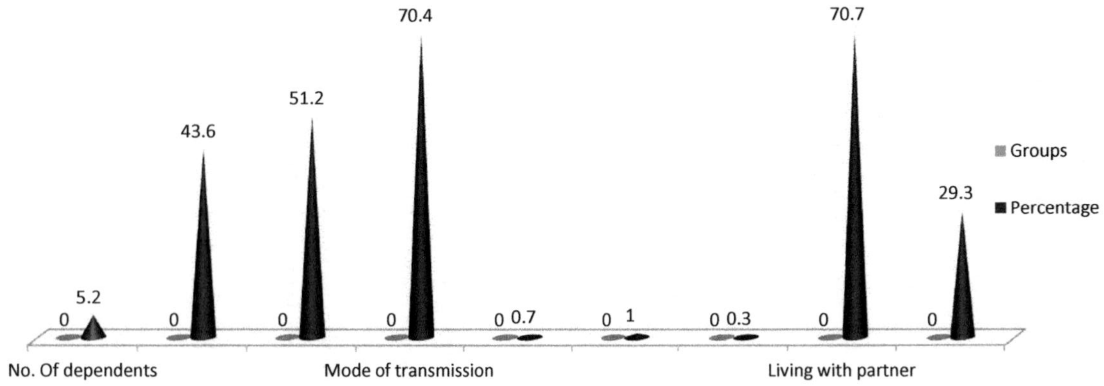

Figure 4: Sample Characteristics according to Number of Dependents, Mode of Transmission and Living with Partner

Table-4

Sample Characteristics according to Disclosure Status, Duration of ART, Family and Social Support

Variable	Total Sample	Groups	Frequency	%
Disclosure Status	287	Partner	107	37.3
		Family	177	61.7
		Friends	1	0.3
		All	2	0.7
Duration of ART	287	More than 5 Years	126	43.9
		Below 5 Years	161	56.1
Family Support	287	Yes	270	94.1
		No	17	5.9
Social Support	287	Yes	199	69.3
		No	88	30.7

Table 4 shows that out of 287 participants, the HIV disclosure status among them, the highest 61.7% (n=177) had disclosed to the family, and remaining 37.3% (n=137) had disclosed to the partner. With regard to duration of ART, majority 56.1% (n=161) were on ART for less than 5 years and remaining 43.9% (n=126) were on ART for more than 5 years. Regarding their family support, the highest 94.9% ((n=270) were having family support and remaining 5.9% (n=7) were not having family support. About their social support 69.3% (n=199) were having social support and remaining 30.7% (n=88) were not having social support.

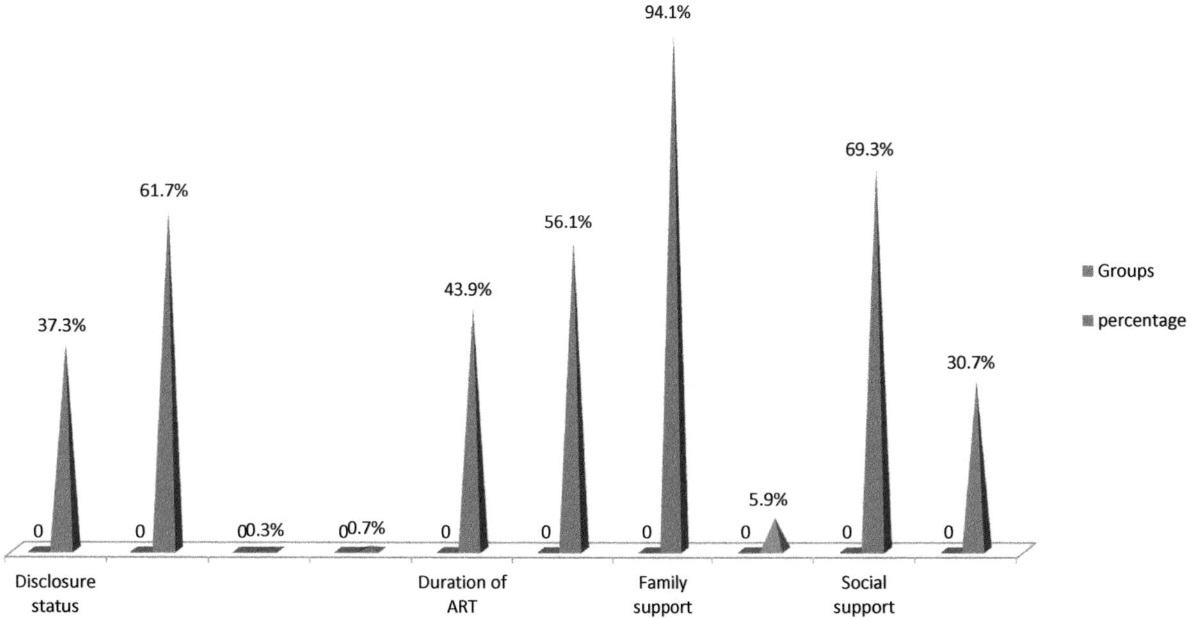

Figure 5: Sample Characteristics according to Disclosure Status, Duration of ART, Family and Social Support

Table -5

Correlation between Appreciation of Music and Emotional & Behavioral Problems of PLHIV on Anti Retroviral Therapy

n=287

Syndrome	**M. Interest**	**M. Aptitude**
Anxious / Depressed	-0.300 (**)	-0.077
Withdrawn	-0.174 (**)	-0.056
Somatic Complaints	-0.043	-0.039
Thought Problems	-0.102	0.107
Attention Problems	-0.241 (**)	0.012
Aggressive Behavior	-0.221 (**)	-0.070
Rule Breaking Behavior	-0.085	0.092
Intrusive	0.014	0.084
Other problems	-0.130 (*)	0.049

** P<0.01 * P< 0.05

Table 5 shows the relationship between Appreciation of music and emotional behavioral problems. A highly significant negative correlation is observed between music interest and anxious, withdrawn, attention problems, aggressive behavior and a significant negative correlation found between music interest and other problems. There is a negative correlation found between music interest and Somatic Complaints, thought problems. From the findings it is understood that the higher the music interest the lesser the emotional and behavioral problems. It is true that the music interest makes the person mentally calm and relaxed and they could exhibit less emotional and behavioral problems. Regarding the relationship between music aptitude and emotional and behavioral problems, there is no significant relationship found between them. This shows the music aptitude has no influence upon emotional and behavioral problems.

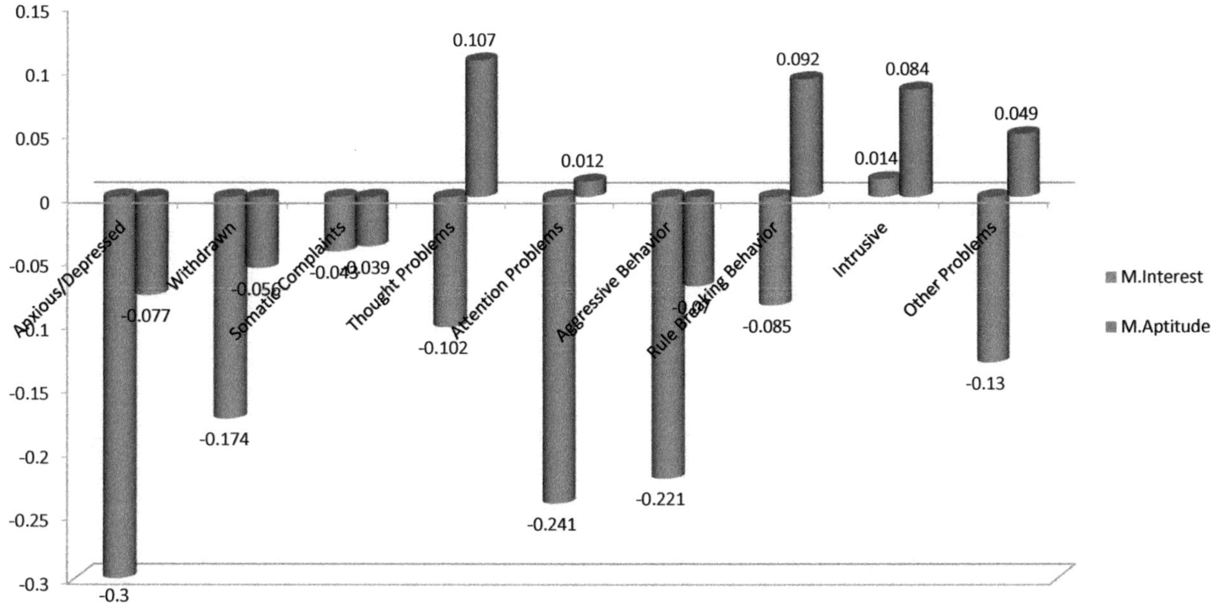

Figure 6: Correlation between Appreciation of Music and Emotional & Behavioral Problems of PLHIV on Anti Retroviral Therapy.

Table-6

Correlation between Appreciation of Music and Wellbeing of PLHIV on Anti Retroviral Therapy

n=287

Subscale	**M_Interest**	**M_Aptitude**
Depression	-0.314(**)	-0.180(**)
Anxiety	-0.239(**)	-0.074
Energy	0.210(**)	0.077
Positive Wellbeing	0.317(**)	0.107
Total Wellbeing	0.344(**)	0.137(*)

** P < 0.01 *P < 0.05

.

Table 6 shows the relation between Appreciation of music and wellbeing. There is a highly significant negative correlation is found between music interest and depression as well as anxiety. It indicates that higher the music interest, the lesser the depression and anxiety.

Very interestingly there is a highly significant positive correlation is noted between music interest and energy, positive wellbeing and total wellbeing. Regarding correlation between music aptitude and wellbeing, there is a highly significant negative correlation observed between music aptitude and depression and in contrast a significant positive correlation observed between music aptitude and total wellbeing. From the findings it is learnt that the higher the music interest, the higher the energy, positive wellbeing and also the total wellbeing. It is a fact that the music promotes wellness feeling and alleviates all the distress. This makes the person more energetic and gets the feeling of total wellbeing.

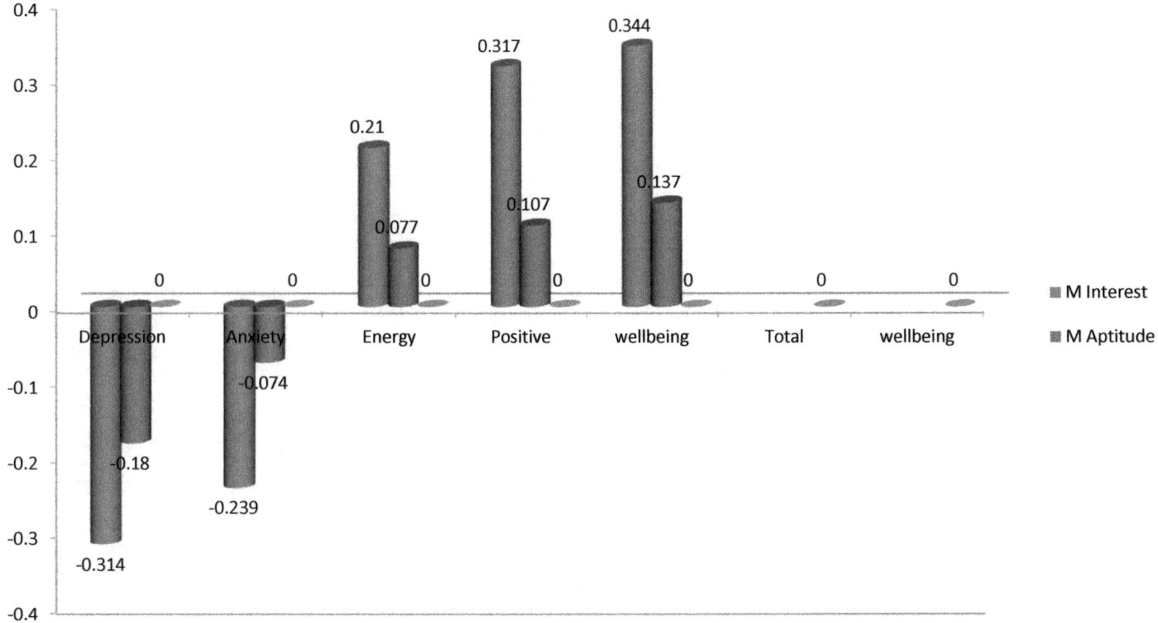

Figure 7 : Correlation between Appreciation of Music and Wellbeing of PLHIV on Anti Retroviral Therapy.

Table-7

Correlation between wellbeing and emotional & behavioral problems of PLHIV on Anti Retroviral Therapy

n=287

Wellbeing Subscale	Anxious / Depressed	Withdrawn	Somatic Complaints	Thought Problems
Depression	0.362(**)	0.156(**)	-0.010	0.098
Anxiety	0.423(**)	0.266(**)	0.122(*)	0.121(*)
Energy	-0.237(**)	-0.228(**)	-0.158(**)	-0.122(*)
Positive Wellbeing	-0.583(**)	-0.405(**)	-0.149(*)	-0.191(**)
Total Wellbeing	-0.529(**)	-0.348(**)	-0.133(*)	-0.172(**)

Contd....

Wellbeing Subscale	Attention Problems	Aggressive Behavior	Rule Breaking Behavior	Intrusive	Other Problems
Depression	0.367(**)	0.193(**)	-0.012	-0.046	0.274(**)
Anxiety	0.512(**)	0.315(**)	0.010	-0.018	0.399(**)
Energy	-0.399(**)	-0.215(**)	-0.025	0.001	-0.341(**)
Positive Wellbeing	-0.680(**)	-0.416(**)	-0.164(**)	-0.063	-0.488(**)
Total Wellbeing	-0.635(**)	-0.373(**)	-0.076	-0.008	-0.481(**)

** P < 0.01 * P < 0.05

Table 7 shows the correlation between wellbeing and emotional behavioural problems. There is a significant positive relationship of wellbeing found with anxious, with drawn, Somatic Complaints and thought problems. Specific to subscales in wellbeing, a highly significant positive correlation of depression found with anxious and withdrawn, and a significant positive correlation observed between anxiety and Somatic Complaints, thought problems. In relation to energy, positive wellbeing and total wellbeing, a highly negative correlation is observed with anxious, withdrawn and thought problems and there is a significant negative correlation observed with somatic complaints.

Above table shows that there is a highly significant positive relationship found between anxiety, depression subscales of wellbeing and attention, aggressive behavior and other problems. In relation to energy, positive wellbeing and total wellbeing, a highly significant negative correlation is found with attention problems, aggressive behavior and other problems. In addition, a highly significant negative correlation found between positive wellbeing and rule breaking behavior. From this findings, it is understood, the higher the depression and anxiety subscales of wellbeing, the higher the emotional and behavioral problems. In contrast, the higher the wellbeing, the lesser the emotional and behavioral problems are observed. The development of emotional and behavioral problems could be due to the chronic state of the disease, side effects, anti retroviral drugs and social stigma.

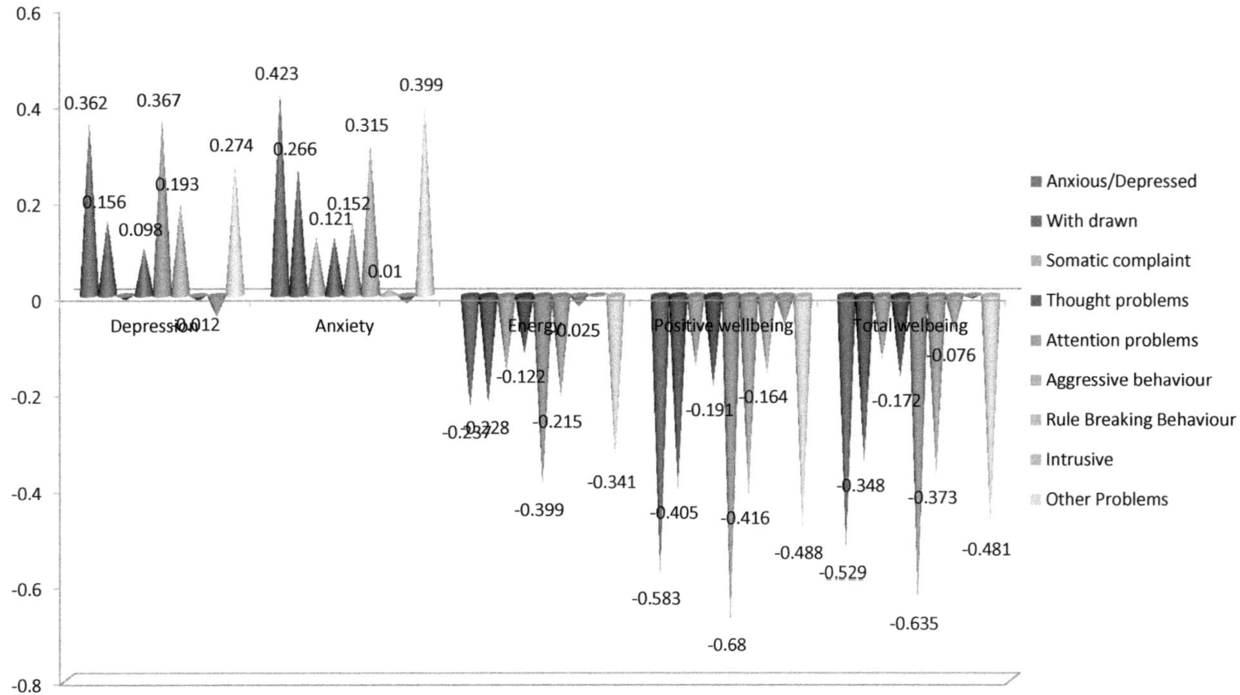

Figure 8 : Correlation between wellbeing and emotional & behavioral problems of PLHIV on Anti Retroviral Therapy.

Table-8

Music interest and Emotional & Behavioral problems of PLHIV on Anti Retroviral Therapy of PLHIV

n = 287

Syndrome	M._Interest	Frequency	Mean	Std. Deviation	t
Anxious / Depressed	Low Interest	85	2.55	1.880	4.543**
	High Interest	202	1.50	1.597	
Withdrawn	Low Interest	85	1.68	1.521	3.190**
	High Interest	202	1.07	1.357	
Somatic Complaints	Low Interest	85	0.60	1.104	0.403
	High Interest	202	0.54	1.282	
Thought Problems	Low Interest	85	0.54	0.880	1.8
	High Interest	202	0.33	0.837	
Attention Problems	Low Interest	85	1.89	1.690	3.113**
	High Interest	202	1.21	1.738	
Aggressive Behavior	Low Interest	85	1.04	1.358	3.192**
	High Interest	202	0.52	0.942	
Rule Breaking Behavior	Low Interest	85	0.44	0.794	1.388
	High Interest	202	0.30	0.713	
Intrusive	Low Interest	85	0.21	0.537	0.247
	High Interest	202	0.19	0.689	
Other problems	Low Interest	85	2.87	2.028	2.133*
	High Interest	202	2.30	2.140	

** $P < 0.01$ * $P < 0.05$

Table 8 shows the significant difference between music interest and emotional behavioral problems and wellbeing. There is a significant difference between emotional problems of low interest and high interest on music. The mean value of low interest on music is high with regard to anxious, withdrawn, attention, aggressive, and other problems. Though there is no significance difference between low interest and high interest on music with regard to somatic complaints, thought problems, rule breaking and intrusive behavior, the mean value of low interest on music is higher than high interest on music. This indicates that the higher the music interest, the lesser the emotional problems among PLHIV.

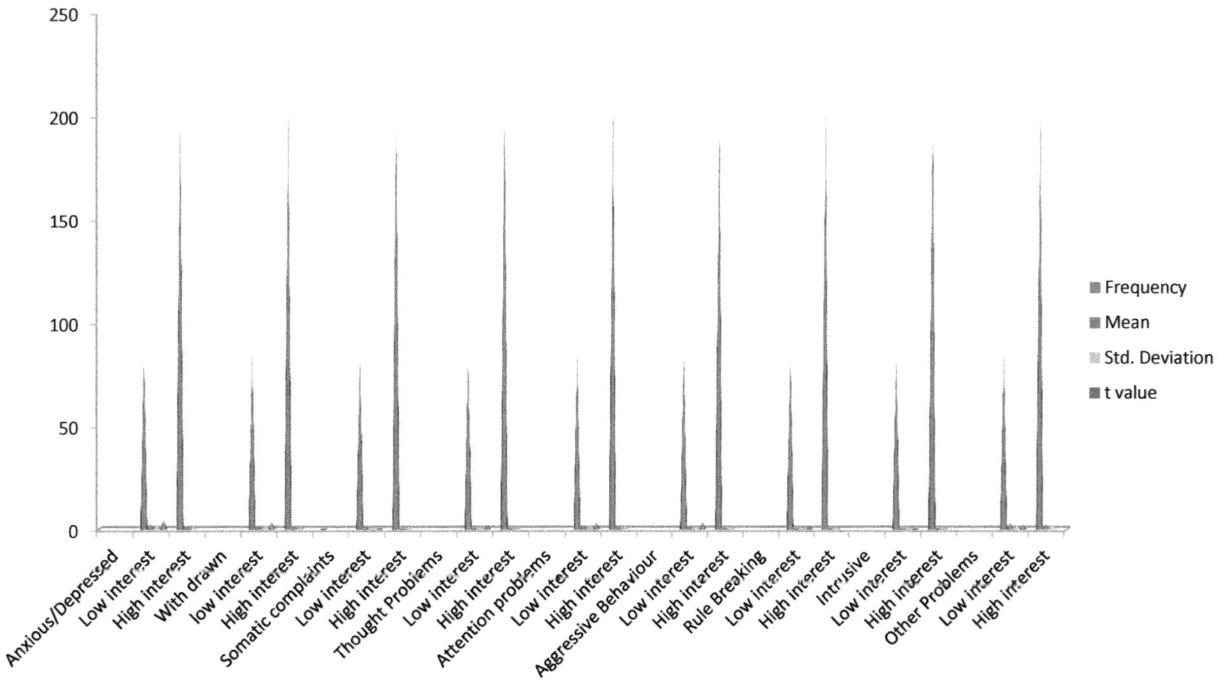

Figure 9 : Music interest and Emotional & Behavioral problems of PLHIV on Anti Retroviral Therapy.

Table-9

Music interest and Wellbeing of PLHIV on Anti Retroviral Therapy

n = 287

Wellbeing Sub scale	M._Interest	Frequency	Mean	Std. Deviation	t
Depression	Low Interest	85	5.62	2.730	4.331**
	High Interest	202	4.12	2.583	
Anxiety	Low Interest	85	5.53	2.113	2.150*
	High Interest	202	4.89	2.682	
Energy	Low Interest	85	9.07	2.120	-1.831
	High Interest	202	9.58	2.282	
Positive Wellbeing	Low Interest	85	8.94	3.837	-3.524**
	High Interest	202	10.71	4.008	
Total Wellbeing	Low Interest	85	42.86	7.889	-4.062**
	High Interest	202	47.29	9.601	

** P <0.01 * P < 0.05

Table 9 shows the significant difference between music interest and wellbeing, there is a significant difference between wellbeing of low interest and high interest on music among PLHIV. The mean value of low interest on music is high with regard to depression ($t=4.331$, $p<0.01$) and anxiety ($t=2.150$, $p<0.05$). This shows that depression is high in low interest on music. Positive wellbeing and total wellbeing is high in high interest on music than low interest on music as the mean value of high interest on music is higher than the low interest on music, Though there is no significant difference is observed between low interest and high interest on music in relation to wellbeing, energy, the mean value is high in high interest on music .This indicates that the wellbeing is high in high interest on music among PLHIV. It is true that music alleviates stress and tension and makes the people free from emotional behavioral problems and thereby enhances the wellbeing.

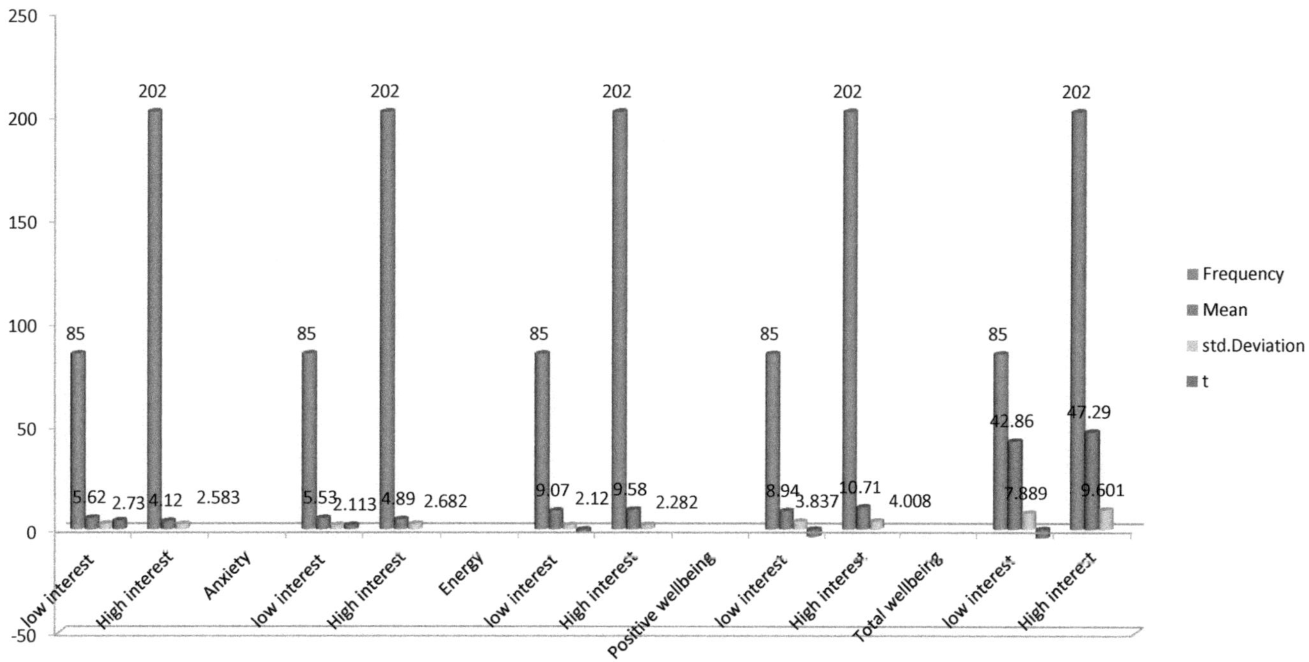

Figure -10: Music interest and Wellbeing of PLHIV on Anti Retroviral Therapy.

Table-10

Music Aptitude and Emotional & Behavioral Problems of PLHIV on Anti Retroviral Therapy

n=287

Syndrome	M.Aptitude	N	Mean	Std. Deviation	t
Anxious / Depressed	Low Aptitude	137	1.85	1.667	0.356
	High Aptitude	150	1.77	1.829	
Withdrawn	Low Aptitude	137	1.31	1.427	0.590
	High Aptitude	150	1.21	1.439	
Somatic Complaints	Low Aptitude	137	0.54	1.176	-0.229
	High Aptitude	150	0.57	1.281	
Thought Problems	Low Aptitude	137	0.31	0.694	-1.540
	High Aptitude	150	0.47	0.974	
Attention Problems	Low Aptitude	137	1.34	1.678	-0.632
	High Aptitude	150	1.47	1.816	
Aggressive Behavior	Low Aptitude	137	0.65	1.068	-0.335
	High Aptitude	150	0.69	1.141	
Rule Breaking Behavior	Low Aptitude	137	0.27	0.624	-1.509
	High Aptitude	150	0.40	0.827	
Intrusive	Low Aptitude	137	0.13	0.497	-1.718
	High Aptitude	150	0.26	0.755	
Other problems	Low Aptitude	137	2.28	1.959	-1.430
	High Aptitude	150	2.64	2.250	

** $P < 0.01$ * $P < 0.05$

Table 10 shows the significant difference between music aptitude and emotional and behavioral problems. There is no significant relationship found between them. Tough there is no significant difference found between them the mean value of low music aptitude high on anxiety and withdrawn. In contrast the mean value of high music aptitude is high on somatic complaints, thought problems, attention problems, aggressive behaviour, rule breaking behavior, intrusive and other problems. This indicates that the music aptitude has got no influence upon emotional and behavioral problems.

Table-11

Music Aptitude and Wellbeing of PLHIV on Anti Retroviral Therapy

n=287

Subscale	M_Aptitude	N	Mean	Std. Deviation	t
Depression	Low Aptitude	137	4.72	2.316	0.956
	High Aptitude	150	4.42	3.028	
Anxiety	Low Aptitude	137	4.96	2.102	-0.753
	High Aptitude	150	5.19	2.886	
Energy	Low Aptitude	137	9.39	1.964	-0.329
	High Aptitude	150	9.47	2.479	
Positive Wellbeing	Low Aptitude	137	10.29	3.413	0.421
	High Aptitude	150	10.09	4.537	
Total Wellbeing	Low Aptitude	137	45.99	7.128	0.030
	High Aptitude	150	45.96	10.997	

** $P < 0.01$ *$P < 0.05$

Table 11 shows that there is no significant difference found between music aptitude and wellbeing. The mean value of low aptitude with reference to depression, positive wellbeing and total wellbeing is higher than the high aptitude. The mean value of high music aptitude in relation to anxiety, energy is higher than the low aptitude. From the findings, it is understood that the aptitude refers to the intellectual ability towards musical performance that may or may not influences the wellbeing.

Table-12

Age and Emotional & Behavioral Problems of PLHIV on Anti Retroviral Therapy

n=287

Syndrome		**N**	**Mean**	**Std. deviation**	**F**
Anxious / Depressed	Below 30 Years	80	1.80	0.244	0.244
	30-40 Years	130	1.75		
	Above 40 Years	77	1.92		
Withdrawn	Below 30 Years	80	1.18	0.169	0.169
	30-40 Years	130	1.28		
	Above 40 Years	77	1.29		
Somatic Complaints	Below 30 Years	80	0.53	0.216	0.216
	30-40 Years	130	0.53		
	Above 40 Years	77	0.64		
Thought Problems	Below 30 Years	80	0.35	0.241	0.241
	30-40 Years	130	0.43		
	Above 40 Years	77	0.38		
Attention Problems	Below 30 Years	80	1.38	0.059	0.059
	30-40 Years	130	1.40		
	Above 40 Years	77	1.47		
Aggressive Behavior	Below 30 Years	80	0.63	0.107	0.107
	30-40 Years	130	0.68		
	Above 40 Years	77	0.70		
Rule Breaking Behavior	Below 30 Years	80	0.18	3.195	3.195*
	30-40 Years	130	0.44		
	Above 40 Years	77	0.34		
Intrusive	Below 30 Years	80	0.23	0.841	0.841
	30-40 Years	130	0.23		
	Above 40 Years	77	0.12		
Other Problems	Below 30 Years	80	2.30	0.369	0.369
	30-40 Years	130	2.55		
	Above 40 Years	77	2.51		

** $P < 0.01$ * $P < 0.05$

Table 12 shows the significant difference between age and emotional behavioral problems. There is no significant difference of age observed with emotional behavioral problems exception rule breaking behavior. There is a significant difference found between the age and rule breaking behavior ($F=3.195$, $p<0.05$). The mean value of 30-40 years is high on rule breaking behavior. Although there is no significant difference found between age and other emotional behavioral problems, the mean value of above 40 years is higher on anxious, withdrawn, somatic complaints, attention problems and aggressive behavior.

Table-13

Age and Wellbeing of PLHIV on Anti Retroviral Therapy

n=287

Wellbeing Sub scale	Age	N	Mean	Std. Deviation	F
Depression	Below 30 Years	80	4.55	0.013	0.013
	30-40 Years	130	4.59		
	Above 40 Years	77	4.53		
Anxiety	Below 30 Years	80	5.41	0.952	0.952
	30-40 Years	130	4.96		
	Above 40 Years	77	4.94		
Energy	Below 30 Years	80	9.36	0.820	0.820
	30-40 Years	130	9.61		
	Above 40 Years	77	9.21		
Positive Wellbeing	Below 30 Years	80	10.05	0.066	0.066
	30-40 Years	130	10.23		
	Above 40 Years	77	10.26		
Total Wellbeing	Below 30 Years	80	45.45	0.197	0.197
	30-40 Years	130	46.28		
	Above 40 Years	77	46.00		

** $p < 0.01$ *$P < 0.05$

Table 13 shows that there is no significant difference found between age and wellbeing. Though there is no significant association found between them, the mean value of 30-40 years is high on depression and the mean value of below 30 years is high on anxiety. Further the mean value of 30-40 years is high on energy and positive wellbeing. It is also observed that the mean value of above 40 years is high on positive wellbeing. However the result did not show any age related differences in emotional behavioral problems as well as in wellbeing.

Table-14

Gender and Emotional & Behavioral Problems of PLHIV on Anti Retroviral Therapy

n=287

Syndrome	Gender	N	Mean	Std. Deviation	t
Anxious / Depressed	Male	176	1.89	1.787	1.033
	Female	111	1.68	1.690	
Withdrawn	Male	176	1.28	1.538	0.463
	Female	111	1.21	1.251	
Somatic Complaints	Male	176	0.55	1.304	-0.114
	Female	111	0.57	1.109	
Thought Problems	Male	176	0.45	0.949	1.488
	Female	111	0.31	0.671	
Attention Problems	Male	176	1.49	1.773	1.022
	Female	111	1.28	1.712	
Aggressive Behavior	Male	176	0.70	1.149	0.521
	Female	111	0.63	1.035	
Rule Breaking Behavior	Male	176	0.45	0.874	3.786**
	Female	111	0.16	0.394	
Intrusive	Male	176	0.22	0.732	0 .623
	Female	111	0.17	0.484	
Other Problems	Male	176	2.59	2.142	1.164
	Female	111	2.29	2.082	

** $P < 0.01$ * $P < 0.05$

Table 14 shows that there is a highly significant difference is observed between gender and rule breaking behavior ($t=3.786$, $p<0.05$). The mean value of males is high on rule breaking behavior. It is also noted that the mean value of male is high on anxious, withdrawn, thought problems; attention problems, aggressive behavior, intrusive and other problems. Generally males exhibit more externalizing problems than women.

Table-15

Gender and Wellbeing of PLHIV on Anti Retroviral Therapy

n=287

Wellbeing subscale	Gender	N	Mean	Std. Deviation	F
Depression	Male	176	4.21	2.775	-2.883*
	Female	111	5.13	2.519	
Anxiety	Male	176	4.93	2.723	-1.306
	Female	111	5.32	2.211	
Energy	Male	176	9.60	2.245	1.571
	Female	111	9.17	2.227	
Positive Wellbeing	Male	176	10.29	4.168	0 .547
	Female	111	10.03	3.824	
Total Wellbeing	MALE	176	46.74	9.780	1.820
	FEMALE	111	44.76	8.487	

**P < 0.01 *P < 0.05

Table 15 shows the significant difference between gender and wellbeing. There is a significant difference is found between gender and depression (F=2.883, p<0.01). The mean value of female is high on depression and anxiety. Further the mean value of male is high on energy, positive wellbeing and total wellbeing. This could be probably due to that males exhibit more subjective wellbeing than females.

Table-16

Education and Emotional Behavioral Problems of PLHIV on Anti Retroviral Therapy

n=287

Syndrome	Education	N	Mean	Std. Deviation	F
Anxious / Depressed	Illiterate	100	1.94	1.874	0.809
	Tenth and Below	144	1.80	1.716	
	Inter and Graduation	43	1.53	1.564	
Withdrawn	Illiterate	100	1.19	1.300	2.413
	Tenth and Below	144	1.41	1.584	
	Inter and Graduation	43	0.88	1.096	
Somatic Complaints	Illiterate	100	0.50	0.905	0.219
	Tenth and Below	144	0.60	1.410	
	Inter and Graduation	43	0.53	1.260	
Thought Problems	Illiterate	100	0.39	0.777	0.329
	Tenth and Below	144	0.37	0.937	
	Inter and Graduation	43	0.49	0.736	
Attention Problems	Illiterate	100	1.44	1.719	0.144
	Tenth and Below	144	1.43	1.792	
	Inter and Graduation	43	1.28	1.709	

Contd-

Aggressive Behavior	Illiterate	100	0.67	1.129	1.783
	Tenth and Below	144	0.76	1.166	
	Inter and Graduation	43	0.40	0.760	
Rule Breaking Behavior	Illiterate	100	0.34	0.728	0.063
	Tenth and Below	144	0.33	0.718.	
	Inter and Graduation	43	0.37	0.846	
Intrusive	Illiterate	100	0.11	0.373	1.914
	Tenth and Below	144	0.27	0.821	
	Inter and Graduation	43	0.16	0.433	
Other Problems	Illiterate	100	2.29	2.001	3.515*
	Tenth and Below	144	2.77	2.286	
	Inter and Graduation	43	1.88	1.621	

** P < 0.01 * P < 0.05

Table 16 shows the significant difference between education and emotional behavioral problems. There is a significant difference observed between education and other problems (F=3.515, p<0.01). The mean value of 10th and below is high on other problems. Although there is no significant difference found between education and other emotional behavioral problems, the mean value of illiterates is high on anxious and attention problems. It is also observed that the mean value of 10th and below is high on withdrawn, somatic complaints, aggressive and intrusive behavior and the mean value of inter and graduation is high on thought problems and rule breaking behavior. These problems could be due to inadequate knowledge on disease condition as majority of PLHIV studied up to 10th standard.

Table-17

Education and Wellbeing of PLHIV on Anti Retroviral Therapy

n=287

Wellbeing subscale	Education	N	Mean	Std. deviation	F
Depression	Illiterate	100	4.88	2.641	1.058
	Tenth and Below	144	4.38	2.681	
	Inter and Graduation	43	4.47	2.963	
Anxiety	Illiterate	100	5.13	2.330	0.812
	Tenth and Below	144	5.18	2.622	
	Inter and Graduation	43	4.63	2.734	
Energy	Illiterate	100	9.30	2.130	1.644
	Tenth and Below	144	9.35	2.295	
	Inter and Graduation	43	10.00	2.299	
Positive Wellbeing	Illiterate	100	10.52	3.907	2.137
	Tenth and Below	144	9.72	4.011	
	Inter and Graduation	43	10.98	4.290	
Total Wellbeing	Illiterate	100	45.81	8.667	1.085
	Tenth and Below	144	45.52	9.432	
	Inter and Graduation	43	47.88	10.443	

** $P < 0.01$ *$P < 0.05$

Table 17 shows that there is no significant difference found between education and wellbeing. There is no significant difference is found between them. The mean value of illiterates is high on depression and the mean value of 10th and below is high on anxiety. It is also noted that the mean value of inter and graduation is high on energy, positive wellbeing and total wellbeing.

Table-18

Occupation and Emotional & Behavioral Problems of PLHIV on Anti Retroviral Therapy

n=287

Syndrome	Occupation	N	Mean	Std. Deviation	F
Anxious / Depressed	Self Employee	41	2.17	2.060	4.613**
	Laborer	69	2.52	1.937	
	Employee	59	1.83	1.631	
	House Wife	54	1.19	1.506	
	Business	21	1.10	1.300	
	Farmer	22	1.36	1.177	
	Grade-IV Employee	21	1.48	1.327	
Withdrawn	Self Employee	41	1.54	1.719	2.280*
	Laborer	69	1.45	1.290	
	Employee	59	1.15	1.311	
	House Wife	54	1.04	1.213	
	Business	21	0.52	0.873	
	Farmer	22	1.09	1.601	
	Grade-IV Employee	21	1.81	2.015	
Somatic Complaints	Self Employee	41	0.39	0.997	1.734
	Laborer	69	0.54	0.994	
	Employee	59	0.59	1.261	
	House Wife	54	0.59	1.141	
	Business	21	0.38	0.805	
	Farmer	22	0.23	0.528	
	Grade-IV Employee	21	1.29	2.552	

Contd.

	Self Employee	41	.71	1.553	2.040
	Laborer	69	0.36	0.707	
	Employee	59	0.47	0.774	
	House Wife	54	0.15	0.359	
	Business	21	0.24	0.436	
	Farmer	22	0.32	0.568	
	Grade-IV Employee	21	0.52	0.928	
Attention Problems	Self Employee	41	1.63	1.933	4.792**
	Laborer	69	2.07	1.726	
	Employee	59	1.68	1.969	
	House Wife	54	0.93	1.528	
	Business	21	0.62	1.203	
	Farmer	22	0.59	1.054	
	Grade-IV Employee	21	0.95	1.431	
Aggressive Behavior	Self Employee	41	0.88	1.418	1.204
	Laborer	69	0.86	1.216	
	Employee	59	0.66	1.044	
	House Wife	54	0.56	0.925	
	Business	21	0.33	0.577	
	Farmer	22	0.41	0.666	
	Grade-IV Employee	21	0.62	1.322	
Rule Breaking Behavior	Self Employee	41	0.44	1.026	1.131
	Laborer	69	0.35	0.682	
	Employee	59	0.32	0.730	
	House Wife	54	0.13	0.339	
	Business	21	0.48	0.928	
	Farmer	22	0.41	0.854	
	Grade-IV	21	0.48	0.680	
	Self Employee	41	0.39	1.202	
	Laborer	69	0.12	0.404	

Contd.

Intrusive	Employee	59	0.24	0.625	1.065
	House Wife	54	0.17	0.505	
	Business	21	0.19	0.512	
	Farmer	22	0.05	0.213	
	Grade-IV Employee	21	0.24	0.539	
Other Problems	Self Employee	41	2.71	2.283	1.223
	Laborer	69	2.91	2.318	
	Employee	59	2.32	2.080	
	House Wife	54	2.11	1.723	
	Business	21	2.19	2.112	
	Farmer	22	1.95	1.558	
	Grade-IV Employee	21	2.71	2.552	

** P < 0.01 * P < 0.05

Table 18 shows the significant difference between occupation and emotional behavioral problems. There is a highly significant difference found between anxiety and occupation (F=4.613, p<0.01). The mean value of laborer is high on anxiety. A similar trend is observed between occupation and attention problems (F=4.792, p<0.01). The mean value of laborer is high on attention problems. Further a significant difference is also observed between occupation and withdrawn (F=2.280, p<0.01)>. The mean of self employee is seen high on withdrawn. This may be due to the influence of work pressures upon emotional health of PLHIV.

Table-19

Occupation and Wellbeing of PLHIV on Anti Retroviral Therapy

n=287

Wellbeing sub scale	Occupation	N	Mean	Std. Deviation	F
Depression	Self Employee	41	3.59	2.439	5.954**
	Laborer	69	5.94	2.999	
	Employee	59	4.73	2.827	
	House Wife	54	4.52	2.178	
	Business	21	3.33	2.331	
	Farmer	22	4.00	2.488	
	Grade-IV Employee	21	3.43	1.748	
Anxiety	Self Employee	41	4.61	2.645	3.900**
	Laborer	69	6.09	2.362	
	Employee	59	5.27	2.803	
	House Wife	54	5.07	2.256	
	Business	21	3.86	2.197	
	Farmer	22	4.36	2.258	
	Grade-IV Employee	21	4.14	2.393	
Energy	Self Employee	41	9.76	1.997	**4.046** **
	Laborer	69	8.62	2.371	
	Employee	59	9.22	2.653	
	House Wife	54	9.30	1.910	
	Business	21	10.52	1.662	
	Farmer	22	10.23	1.850	
	Grade-IV Employee	21	10.48	1.662	

Contd-

Positive Wellbeing	Self Employee	41	10.66	3.991	5.182**
	Laborer	69	8.42	3.533	
	Employee	59	9.58	4.304	
	House Wife	54	10.72	3.698	
	Business	21	12.43	3.627	
	Farmer	22	12.00	3.162	
	Grade-IV Employee	21	11.29	4.638	
Total Wellbeing	Self Employee	41	48.22	8.825	7.386**
	Laborer	69	41.01	8.524	
	Employee	59	44.80	10.483	
	House Wife	54	46.43	8.004	
	Business	21	51.76	8.336	
	Farmer	22	49.86	7.485	
	Grade-IV Employee	21	50.19	7.521	

**= $p<0.01$, *$p<0.05$

Table 19 shows the significant difference between occupation and wellbeing. There is a highly significant difference is observed between occupation and depression($F=5.954$, $p<0.01$), anxiety($F=3.9$, $p<0.01$), energy($F=4.046$, $p<0.01$), positive wellbeing ($F=5.182$, $p<0.01$) and total wellbeing($F=7.386$, $p<0.01$). The mean value of laborer is high on depression as well as anxiety. This could be due to the influence of sickness on daily wage earning leading to anxiety and depression among PLHIV. It is also found that the mean value of business is high on energy, positive wellbeing and also on total wellbeing. This is probably due to the flexible working hours and economic security might be the leading factors for the development of positive wellbeing.

Table-20

Area of living and Emotional & Behavioral Problems of PLHIV on Anti Retroviral Therapy

n=287

Syndrome	Area of living	N	Mean	Std. Deviation	F
Anxious / Depressed	Urban	233	1.84	1.774	0.329
	Rural	54	1.69	1.658	
Withdrawn	Urban	233	1.28	1.413	0.504
	Rural	54	1.13	1.518	
Somatic Complaints	Urban	233	0.57	1.262	0.067
	Rural	54	0.52	1.094	
Thought Problems	Urban	233	0.39	0.712	0.017
	Rural	54	0.41	1.311	
Attention Problems	Urban	233	1.40	1.759	0.024
	Rural	54	1.44	1.723	
Aggressive Behavior	Urban	233	0.62	1.040	3.028
	Rural	54	0.91	1.336	
Rule Breaking Behavior	Urban	233	0.36	0.770	0.754
	Rural	54	0.26	0.589	
Intrusive	Urban	233	0.21	0.550	0.162
	Rural	54	0.17	0.966	
Other Problems	Urban	233	2.45	2.115	0.066
	Rural	54	2.54	2.161	

** $P < 0.01$ * $P < 0.05$

Table 20 shows the significant difference between area of living and emotional behavioral problems. Although there is no significant association found between area of living and emotional behavioral problems, the mean value of urban area of living is high on anxiety, withdrawn, somatic complaints, rule breaking and intrusive behavior. The reasons could be due to various tensions associated with the urban life style, sickness might be the influencing factors to cause emotional behavioral problems.

Table-21

Area of living and Wellbeing of PLHIV on Anti Retroviral Therapy

n=287

Wellbeing subscale	Area of living	N	Mean	Std. deviation	F
Depression	Urban	233	4.51	2.698	0.566
	Rural	54	4.81	2.782	
Anxiety	Urban	233	5.14	2.620	0.725
	Rural	54	4.81	2.164	
Energy	Urban	233	9.44	2.193	0.008
	Rural	54	9.41	2.476	
Positive Wellbeing	Urban	233	10.24	4.091	0.207
	Rural	54	9.96	3.806	
Total Wellbeing	Urban	233	46.03	9.486	0.042
	Rural	54	45.74	8.742	

** $P < 0.01$ * $P < 0.05$

Table 21 shows that there is no significant difference found between area of living and wellbeing, but the mean value of urban area of living is high on anxiety, energy, positive wellbeing and total wellbeing. This might be due to availability of health services, educational and divisional facilities in urban area.

Table-22

Marital Status and Emotional & Behavioral Problems of PLHIV on Anti Retroviral Therapy

n=287

Syndrome	Marital Status	N	Mean	Std. Deviation	F
Anxious / Depressed	Married	230	1.71	1.737	1.728
	Single	18	2.17	1.724	
	Married but Single	39	2.21	1.809	
Withdrawn	Married	230	1.17	1.318	2.580
	Single	18	1.83	2.176	
	Married but Single	39	1.51	1.604	
Somatic Complaints	Married	230	0.55	1.224	0.078
	Single	18	0.67	1.534	
	Married but Single	39	0.56	1.142	
Thought Problems	Married	230	0.31	0.611	8.348**
	Single	18	1.11	2.193	
	Married but Single	39	0.54	0.884	
Attention Problems	Married	230	1.30	1.706	2.410
	Single	18	1.61	2.004	
	Married but Single	39	1.95	1.820	
Aggressive Behavior	Married	230	0.62	1.062	1.470
	Single	18	0.94	1.392	
	Married but Single	39	0.87	1.196	
Rule Breaking Behavior	Married	230	0.33	0.751	0.545
	Single	18	0.50	0.707	
	Married but Single	39	0.28	0.686	
Intrusive	Married	230	0.15	0.462	7.047**
	Single	18	0.72	1.674	
	Married but Single	39	0.26	0.677	
Other Problems	Married	230	2.33	1.975	2.750
	Single	18	2.78	2.840	
	Married but Single	39	3.15	2.466	

** $P < 0.01$ * $P < 0.05$

Table 22 shows the significant difference between marital status and emotional behavioral problems, wellbeing. There is a highly significant difference found between marital status and thought problems ($F=8.348$, $p<0.01$) and similarly between marital status and intrusive behavior. The mean value of single is seen high on thought problems and intrusive behavior. Though there is no significant difference is seen with other emotional behavioral problems, the mean value of single is high on withdrawn, thought problems, aggressive behavior, somatic complaints, rule breaking and other problems. The reasons for such difference could be the feeling of loneliness with a chronic disease might be the influencing factor on emotional health of PLHIV.

Table-23

Marital Status and Wellbeing of PLHIV on Anti Retroviral Therapy

n=287

Wellbeing subscale	Marital Status	N	Mean	Sd. Deviation	F
Depression	Married	230	4.42	2.794	2.960*
	Single	18	4.33	2.473	
	Married but Single	39	5.54	2.101	
Anxiety	Married	230	4.96	2.615	1.563
	Single	18	5.28	2.803	
	Married but Single	39	5.72	1.820	
Energy	Married	230	9.57	2.184	2.622
	Single	18	9.28	2.191	
	Married but Single	39	8.69	2.515	
Positive Wellbeing	Married	230	10.52	4.032	5.549**
	Single	18	10.17	4.502	
	Married but Single	39	8.23	3.288	
Total Wellbeing	Married	230	46.72	9.595	5.019**
	Single	18	45.83	9.395	
	Married but Single	39	41.67	6.267	

**= $p<0.01$, *=$p<0.05$

Table 23 shows the significant difference between marital status and wellbeing. There is a highly significant association is found between marital status and positive wellbeing ($F=5.549$, $p<0.01$). A similar trend is observed with total wellbeing ($F=5.019$, $p<0.01$ also. A significant association is also noted between marital status and depression ($F=2.960$, $p<0.05$). The mean value of married but single is high on depression. Though there is no significant difference between marital status and anxiety, depression and energy, the mean value of single is high on anxiety and the mean value of married is high on energy. This could be the fact that married people feel more secure emotionally as they receive support from partner in times of need. As the body and mind integrated together the emotional health influences on physical health and helps to promote wellbeing of PLHIV.

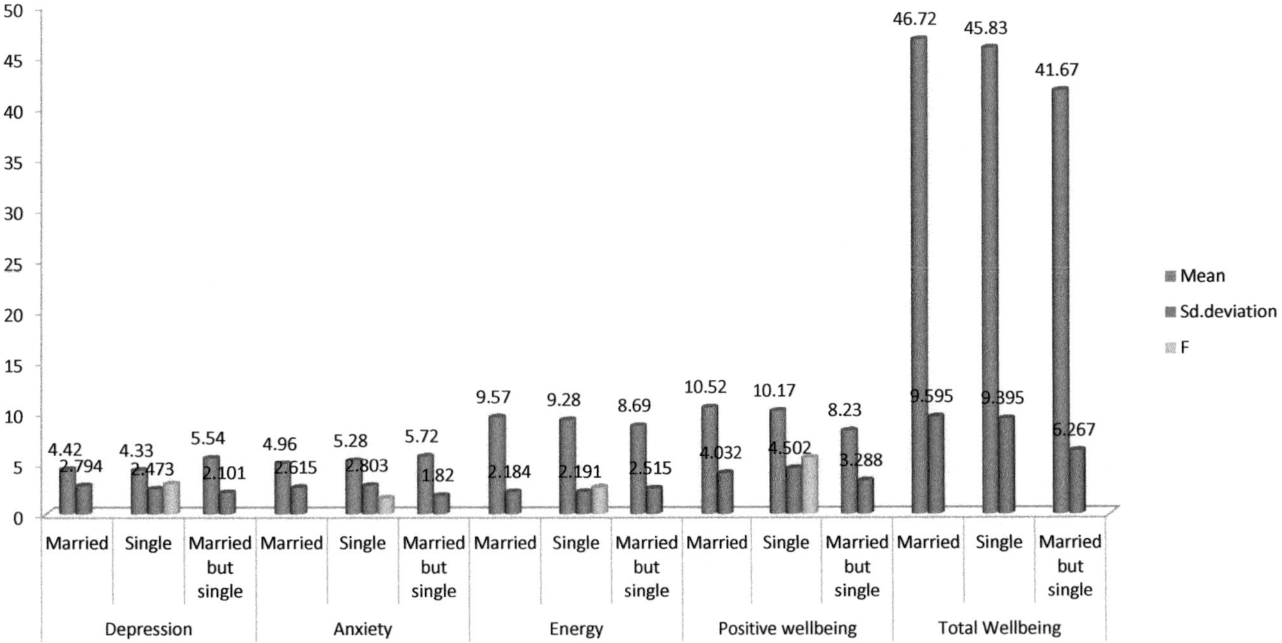

Figure 11: Marital Status and Wellbeing of PLHIV on Anti Retroviral Therapy.

Table-24

Caste and Emotional & Behavioral problems of PLHIV on Anti Retroviral Therapy

n=287

Syndrome	Caste	N	Mean	Std. Deviation	Std. Error	F
Anxious / Depressed	SC&ST	44	1.50	1.691	0.255	2.340
	BC	173	1.99	1.722	0.131	
	OC	70	1.56	1.823	0.218	
Withdrawn	SC&ST	44	1.18	1.244	0.188	0.074
	BC	173	1.26	1.421	0.108	
	OC	70	1.29	1.580	0.189	
Somatic Complaints	SC&ST	44	0.43	1.354	0.204	0.411
	BC	173	0.61	1.265	0.096	
	OC	70	0.51	1.060	0.127	
Thought Problems	SC&ST	44	0.39	0.813	0.123	1.033
	BC	173	0.45	0.936	0.071	
	OC	70	0.27	0.635	0.076	
Attention Problems	SC&ST	44	1.05	1.555	0.234	3.305*
	BC	173	1.62	1.828	0.139	
	OC	70	1.11	1.602	0.191	
Aggressive Behavior	SC&ST	44	0.48	0.821	0.124	0.811
	BC	173	0.71	1.121	0.085	
	OC	70	0.71	1.218	0.146	
Rule Breaking Behavior	SC&ST	44	0.32	0.674	0.102	0.040
	BC	173	0.34	0.741	0.056	
	OC	70	0.36	0.781	0.093	
Intrusive	SC&ST	44	0.14	0.409	0.062	1.317
	BC	173	0.25	0.764	0.058	
	OC	70	0.11	0.401	0.048	
Other problems	SC&ST	44	2.05	1.842	0.278	3.432*
	BC	173	2.73	2.124	0.161	
	OC	70	2.09	2.198	0.263	

** $P < 0.01$ * $P < 0.05$

Table 24 shows the significant difference between caste and emotional problems. There is a significant association found between caste and attention problems ($F=3.305$, $p<0.05$) and similar trend is observed with other problems also ($F=3.432$, $p<0.05$). The mean value of B.C is high on attention problems and other problems. Though there is no significant difference found between caste and anxious, withdrawn, thought problems, attention problems, aggressive and intrusive behavior, the mean value of those are high among B.C. So it could be true that the ignorance, illiteracy, and poverty is high in B.C, that might be the causative factor for development of emotional behavioral problems among them.

Table-25

Caste and Wellbeing of PLHIV on Anti Retroviral Therapy

n=287

Wellbeing subscale	Caste	N	Mean	Std. deviation	Std. error	F
Depression	SC&ST	44	4.43	2.618	0.395	0.114
	BC	173	4.62	2.896	0.220	
	OC	70	4.50	2.295	0.274	
Anxiety	SC&ST	44	4.80	2.825	0.426	1.166
	BC	173	5.27	2.501	0.190	
	OC	70	4.80	2.441	0.292	
Energy	SC&ST	44	9.70	2.064	0.311	2.345
	BC	173	9.20	2.360	0.179	
	OC	70	9.83	2.000	0.239	
Positive Wellbeing	SC&ST	44	10.59	4.337	0.654	0.345
	BC	173	10.05	4.056	0.308	
	OC	70	10.29	3.815	0.456	
Total wellbeing	SC&ST	44	47.07	9.778	1.474	0.962
	BC	173	45.36	9.657	0.734	
	OC	70	46.81	8.174	0.977	

**= $p<0.01$, *=$p<0.05$

Table 25 shows that there is a significant difference between caste and wellbeing. The mean value of SC, ST is found high on positive wellbeing and total wellbeing and the mean value of B.C is high on depression and anxiety. The wellbeing among SC and ST could be due to the welfare measures of government and socio economic empowerment might be the influencing factor on development of wellbeing.

Table-26

Religion and Emotional & Behavioral Problems of PLHIV on Anti Retroviral Therapy

n=287

Syndrome	Religion	N	Mean	Std. Deviation	t
Anxious / Depressed	Hindu	224	1.82	1.724	0.228
	Others	63	1.76	1.855	
Withdrawn	Hindu	224	1.29	1.515	0.837
	Others	63	1.14	1.090	
Somatic Complaints	Hindu	224	0.54	1.193	-0.309
	Others	63	0.60	1.362	
Thought Problems	Hindu	224	0.39	0.891	-0.226
	Others	63	0.41	0.710	
Attention Problems	Hindu	224	1.39	1.688	-0.306
	Others	63	1.48	1.966	
Aggressive	Hindu	224	0.67	1.099	0.046
	Others	63	0.67	1.136	
Rule Breaking Behavior	Hindu	224	0.35	0.761	0.688
	Others	63	0.29	0.658	
Intrusive	Hindu	224	0.21	0.701	1.044
	Others	63	0.14	0.396	
Other problems	Hindu	224	2.47	2.166	0.045
	Others	63	2.46	1.966	

** $P < 0.01$ * $P < 0.05$

Table 26 shows that there is no significant difference between religion and emotional behavioral problems. Although there is no significant difference between them, the mean value of Hindus is high on anxious, withdrawn, rule breaking, intrusive behavior and other problems.

Table-27

Religion and Wellbeing of PLHIV on Anti Retroviral Therapy

n=287

Wellbeing subscale	Religion	N	Mean	Std. deviation	t
Depression	Hindu	224	4.45	2.673	-1.298
	Others	63	4.97	2.828	
Anxiety	Hindu	224	4.93	2.349	-1.646
	Others	63	5.62	3.087	
Energy	Hindu	224	9.50	2.255	0.914
	Others	63	9.21	2.208	
Positive wellbeing	Hindu	224	10.27	4.016	0.622
	Others	63	9.90	4.118	
Total Wellbeing	Hindu	224	46.38	9.064	1.310
	Others	63	44.52	10.191	

** $p<0.01$ * $p<0.05$

Table 27 shows there is no significant difference between religion and wellbeing. The mean value of Hindus is high on energy, positive wellbeing and total wellbeing. It is also noted that the other religions is high on depression and anxiety. This is probably due to cultural diversity and diversified concepts of Hindu might be the root cause for development of emotional behavioral problems as well as wellbeing among them.

Table-28

Income and Emotional & Behavioral Problems of PLHIV on Anti Retroviral Therapy

n=287

Syndrome	Income	N	Mean	Std. Deviation	F
Anxious / Depressed	Below Rs.3000	96	1.97	1.762	3.320*
	Rs.3000-6000	118	1.96	1.790	
	Above Rs.6000	73	1.36	1.610	
Withdrawn	Below Rs.3000	96	1.40	1.559	0.998
	Rs.3000-6000	118	1.25	1.281	
	Above Rs.6000	73	1.08	1.488	
Somatic Complaints	Below Rs.3000	96	0.58	1.262	0.043
	Rs.3000-6000	118	0.53	1.344	
	Above Rs. 6000	73	0.56	0.986	
Thought Problems	Below Rs. 3000	96	0.34	0.708	0.564
	Rs.3000-6000	118	0.46	0.735	
	Above Rs. 6000	73	0.36	1.159	
Attention Problems	Below Rs.3000	96	1.49	1.704	2.458
	Rs.3000-6000	118	1.58	1.827	
	Above Rs.6000	73	1.03	1.641	
Aggressive Behavior	Below Rs.3000	96	0.71	1.015	1.564
	Rs.3000-6000	118	0.76	1.203	
	Above Rs.6000	73	0.48	1.042	
Rule Breaking Behavior	Below Rs.3000	96	0.42	0.735	0.975
	Rs.3000-6000	118	0.32	0.772	
	Above Rs.6000	73	0.26	0.688	
Intrusive	Below Rs.3000	96	0.13	0.391	0.934
	Rs.3000-6000	118	0.24	0.609	
	Above Rs. 6000	73	0.23	0.921	
Other Problems	Below Rs.3000	96	2.46	2.243	1.649
	Rs.3000-6000	118	2.69	2.150	
	Above Rs.6000	73	2.12	1.870	

** P < 0.01 * P < 0.05

Table 28 shows that significant difference between income and emotional behavioral problems. There is a significant difference found between income and anxious ($F=3.320$, $p<0.05$). The mean value of below Rs 3000 is high on anxious. Although there is no significant association of income with other emotional behavioral problems, the mean value of below Rs 3000 is high on withdrawn, somatic complaints and rule breaking behavior. Further the mean value of Rs 3000-6000 is high on thought problems, attention problems, aggressive behavior, intrusive behavior and other problems. The reasons could be due to the inability to meet the economic demands in fulfillment of the individual and family needs due to poor economic status could be the leading cause for development of emotional behavioral problems.

Table-29

Income and Wellbeing of PLHIV on Anti Retroviral Therapy

n=287

Wellbeing subscale	Income	N	Mean	Std. deviation	F
Depression	Below Rs.3000	96	5.15	2.595	3.372*
	Rs.3000-6000	118	4.28	2.828	
	Above Rs.6000	73	4.26	2.577	
Anxiety	Below Rs.3000	96	5.45	2.247	4.129*
	Rs.3000-6000	118	5.22	2.721	
	Above Rs.6000	73	4.37	2.492	
Energy	Below Rs.3000	96	9.26	2.244	0.533
	Rs.3000-6000	118	9.46	2.304	
	Above Rs.6000	73	9.62	2.158	
Positive Wellbeing	Below Rs.3000	96	9.47	3.861	4.003*
	Rs.3000-6000	118	10.14	4.234	
	Above Rs.6000	73	11.22	3.750	
Total Wellbeing	Below Rs.3000	96	44.14	8.173	4.040*
	Rs.3000-6000	118	46.09	9.896	
	Above Rs.6000	73	48.21	9.455	

** $P<0.01$ * $P < 0.05$

Table 29 shows the significant difference between wellbeing and income. There is a significant difference is observed between income and depression ($F=3.372$, $p<0.05$) and anxiety ($F=4.129$, $p<0.05$), energy ($F=4.003$, $p<0.05$), positive wellbeing ($F=4.003$, $p<0.05$) and total wellbeing ($F=4.040$, $p<0.05$). The mean value of below Rs 3000 is high on depression, anxiety and the mean value of above Rs 6000 is high on energy, positive wellbeing and total wellbeing. This indicates that the high income helps in meeting the basic needs of individuals, there by maintains the health of PLHIV resulting in promotion of wellbeing feeling.

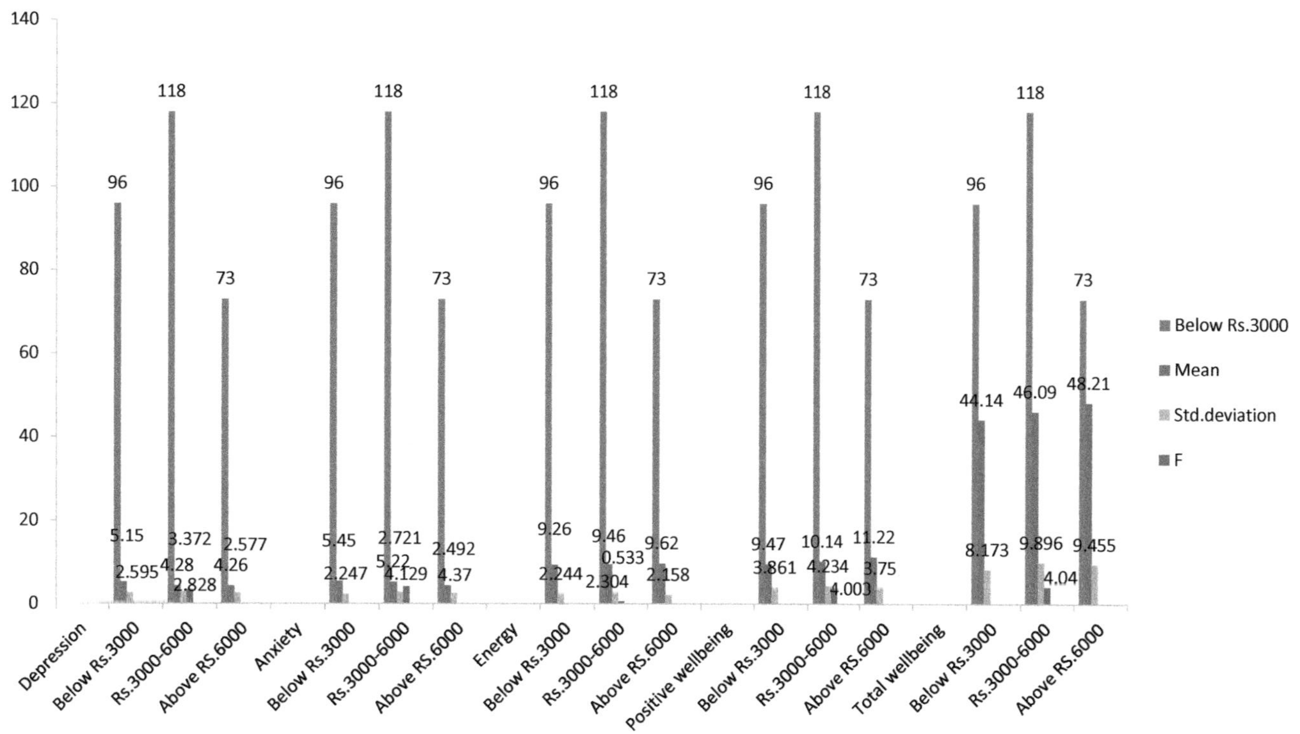

Figure 12: Income and Wellbeing of PLHIV on Anti Retroviral Therapy.

Table-30

No. of Dependents and Emotional & Behavioral Problems of PLHIV on Anti Retroviral Therapy

n=287

Syndrome	No. of dependents	N	Mean	Std. Deviation	F
Anxious / Depressed	No Dependents	15	2.07	1.438	1.373
	One or Two	125	1.62	1.655	
	More than Two	147	1.95	1.850	
Withdrawn	No Dependents	15	1.47	2.326	1.528
	One or Two	125	1.09	1.276	
	More than Two	147	1.37	1.439	
Somatic Complaints	No Dependents	15	0.53	1.356	0.385
	One or Two	125	0.49	0.956	
	More than Two	147	0.62	1.416	
Thought Problems	No Dependents	15	0.73	2.314	2.716
	One or Two	125	0.28	0.576	
	More than Two	147	0.46	0.778	
Attention Problems	No Dependents	15	0.93	1.751	2.334
	One or Two	125	1.22	1.621	
	More than Two	147	1.62	1.837	
Aggressive Behavior	No Dependents	15	0.80	1.373	1.327
	One or Two	125	0.55	0.911	
	More than Two	147	0.76	1.218	
Rule Breaking Behavior	No Dependents	15	0.20	0.414	0.795
	One or Two	125	0.30	0.672	
	More than Two	147	0.39	0.815	
	No Dependents	15	0.73	1.907	
	One or Two	125	0.14	0.416	
	More than Two	147	0.19	0.541	
	No Dependents	15	1.93	2.549	
	One or Two	125	2.29	1.946	
	More than Two	147	2.68	2.205	

Contd..

Intrusive	No Dependents	15	.73	1.907	5.768**
	One or Two	125	.14	.416	
	More than Two	147	.19	.541	
Other Problems	No Dependents	15	1.93	2.549	1.672
	One or Two	125	2.29	1.946	
	More than Two	147	2.68	2.205	

** P < 0.01 * P < 0.05

Table 30 shows that there is a highly significant difference found between no. of dependants and intrusive behavior (F=5.768, p<0.01). The mean value of no dependants is high on intrusive behavior. This indicates, as there are no dependants to share their life could be the influencing factor to develop an intrusive behavior.

Table-31

No. of Dependents and Wellbeing of PLHIV on Anti Retroviral Therapy

n=287

Wellbeing subscale	**No. of Dependents**	**N**	**Mean**	**Std. deviation**	**F**
Depression	No Dependents	15	4.00	2.803	0.554
	One or Two	125	4.71	2.691	
	More than Two	147	4.50	2.728	
Anxiety	No Dependents	15	4.07	1.944	1.267
	One or Two	125	5.12	2.461	
	More than Two	147	5.15	2.649	
Energy	No Dependents	15	9.67	1.496	0.086
	One or Two	125	9.42	2.356	
	More than Two	147	9.41	2.220	
Positive Wellbeing	No Dependents	15	10.60	4.014	0.527
	One or Two	125	10.42	4.018	
	More than Two	147	9.95	4.062	
Total wellbeing	No Dependents	15	48.20	8.445	0.479
	One or Two	125	46.01	9.308	
	More than Two	147	45.72	9.474	

** $P < 0.01$ * $P < 0.05$

Table 31 shows that there is no significant difference found between no. of dependants and wellbeing. The mean value of number of dependants is high on energy, positive wellbeing and total wellbeing. This could probably due to the increased attention towards their own health as there are no dependants might be the causative factor for wellbeing.

Table-32

Living with Partner and Emotional & Behavioral Problems of PLHIV on Anti Retroviral Therapy

n=287

Syndrome	Living with Partner	N	Mean	Std. Deviation	t
Anxious / Depressed	Yes	203	1.83	1.765	0.291
	No	84	1.76	1.726	
Withdrawn	Yes	203	1.23	1.349	-0.472
	No	84	1.32	1.622	
Somatic Complaints	Yes	203	0.58	1.300	0.556
	No	84	0.50	1.047	
Thought Problems	Yes	203	0.34	0.644	-1.191
	No	84	0.51	1.217	
Attention Problems	Yes	203	1.41	1.742	0.039
	No	84	1.40	1.777	
Aggressive Behavior	Yes	203	0.67	1.111	-0.178
	No	84	0.69	1.097	
Rule Breaking Behavior	Yes	203	0.34	0.736	0.068
	No	84	0.33	0.750	
Intrusive	Yes	203	0.17	0.489	-0.999
	No	84	0.27	0.923	
Other Problems	Yes	203	2.45	2.037	-0.260
	No	84	2.52	2.321	

** $P < 0.01$ * $P < 0.05$

Table 32 shows that there is no significant difference found between living with partner and emotional behavioral problem. But the mean value of living with partner is high on anxious, somatic complaints, attention problems and rule breaking behavior and the mean value of not living with partner is high on withdrawn thought problems, aggressive behavior, intrusive behavior and other problems. As the health is an individual experience and sickness due to HIV could be influencing one's emotional health might be the cause for emotional problems, irrespective of living with partner or not.

Table-33

Living with Partner and Wellbeing of PLHIV on Anti Retroviral Therapy

n=287

Wellbeing subscale	Living with partner	N	Mean	Std. deviation	t
Depression	Yes	203	4.48	2.772	-0.869
	No	84	4.77	2.562	
Anxiety	Yes	203	5.15	2.643	0.745
	No	84	4.92	2.277	
Energy	Yes	203	9.51	2.169	0.845
	No	84	9.25	2.419	
Positive wellbeing	Yes	203	10.23	4.092	0.288
	No	84	10.08	3.912	
Total wellbeing	Yes	203	46.11	9.612	0.405
	No	84	45.64	8.680	

** $P < 0.01$ *$P < 0.05$

Table 33 shows that there is no significant difference between living with partner and wellbeing, the mean value of living with partner is high on energy, positive wellbeing and total wellbeing. As living with partner is a source of support for PLHIV might be the influencing factor for the development of mental wellbeing among them.

Table-34

Family Support and Emotional & Behavioral Problems of PLHIV on Anti Retroviral Therapy

n=287

Syndrome	Family Support	N	Mean	Std. Deviation	t
Anxious / Depressed	Yes	270	1.85	1.767	2.178*
	No	17	1.12	1.317	
Withdrawn	Yes	270	1.25	1.380	-0.199
	No	17	1.35	2.149	
Somatic Complaints	Yes	270	0.55	1.233	-0.520
	No	17	0.71	1.213	
Thought Problems	Yes	270	0.37	0.682	-0.720
	No	17	0.76	2.251	
Attention Problems	Yes	270	1.43	1.748	0.552
	No	17	1.18	1.811	
Aggressive Behavior	Yes	270	0.65	1.065	-0.874
	No	17	1.00	1.620	
Rule Breaking Behavior	Yes	270	0.34	0.754	0. 945
	No	17	0.24	0.437	
Intrusive	Yes	270	0.18	0.518	-0.699
	No	17	0.47	1.700	
Other Problems	Yes	270	2.51	2.103	1.176
	No	17	1.82	2.351	

** $P < 0.01$ * $P < 0.05$

Table 34 shows that there is a significant difference between family support and anxious ($t=2.178$, $p<0.05$). The mean value of family support is high on anxious. Though there is no significant difference found between family support and other emotional behavioral problems, the mean value of no family support is high on withdrawn, somatic complaints, thought problems, aggressive and intrusive behavior. This indicates that even though having family support among PLHIV, the anxiety could be due to fatality of the disease might be causing anxious towards outcome of the disease in spite of using ART.

Table-35

of PLHIV on Anti Retroviral Therapy Family Support and Wellbeing

n=287

Wellbeing subscale	Family support	N	Mean	Std. deviation	t
Depression	Yes	270	4.57	2.719	0 .243
	No	17	4.41	2.671	
Anxiety	Yes	270	5.08	2.573	0 .044
	No	17	5.06	2.015	
Energy	Yes	270	9.49	2.223	1.573
	No	17	8.53	2.452	
Positive wellbeing	Yes	270	10.16	4.066	-0.472
	No	17	10.59	3.572	
Total wellbeing	Yes	270	46.00	9.427	0.173
	No	17	45.65	7.984	

** $P < 0.01$ * $P < 0.05$

Table 35 shows that there is no significant difference found between family support and wellbeing. Though there is no significant difference between family support and wellbeing, the mean value of family support is high on energy and total wellbeing. It is true that the family support always enhances the individual feeling of wellbeing.

Table-36

Social Support and Emotional & Behavioral Problems of PLHIV on Anti Retroviral Therapy

n=287

Syndrome	Social Support	N	Mean	Std. Deviation	t
Anxious / Depressed	Yes	199	2.05	1.847	4.024
	No	88	1.26	1.369	
Withdrawn	Yes	199	1.39	1.523	2.642*
	No	88	0.95	1.154	
Somatic Complaints	Yes	199	0.56	1.174	0.006
	No	88	0.56	1.355	
Thought Problems	Yes	199	0.42	0.939	0 .987
	No	88	0.33	0.620	
Attention Problems	Yes	199	1.68	1.852	4.543**
	No	88	0.81	1.312	
Aggressive Behavior	Yes	199	0.74	1.105	1.655
	No	88	0.51	1.093	
Rule Breaking Behavior	Yes	199	0.36	0.798	0 .922
	No	88	0.28	0.586	
Intrusive	Yes	199	0.24	0.733	2.124*
	No	88	0.10	0.373	
Other Problems	Yes	199	2.69	2.202	2.907*
	No	88	1.97	1.835	

** $P < 0.01$ * $P < 0.05$

Table 36 shows that there is a significant difference between social support and with drawn ($t=2.642$, $p<0.05$), intrusive ($t=2.124$, $p<0.05$) and other problems ($t=2.907$, $p<0.05$) and highly significant difference found with attention problems ($t=4.543$, $p<0.01$). From the findings it is observed that the mean value of social support is high on emotional behavioral problems. This could be probably due to the personality differences among PLHIV, opportunistic infections due to HIV, family problems and financial problems causing emotional behavioral problems in spite of having social support.

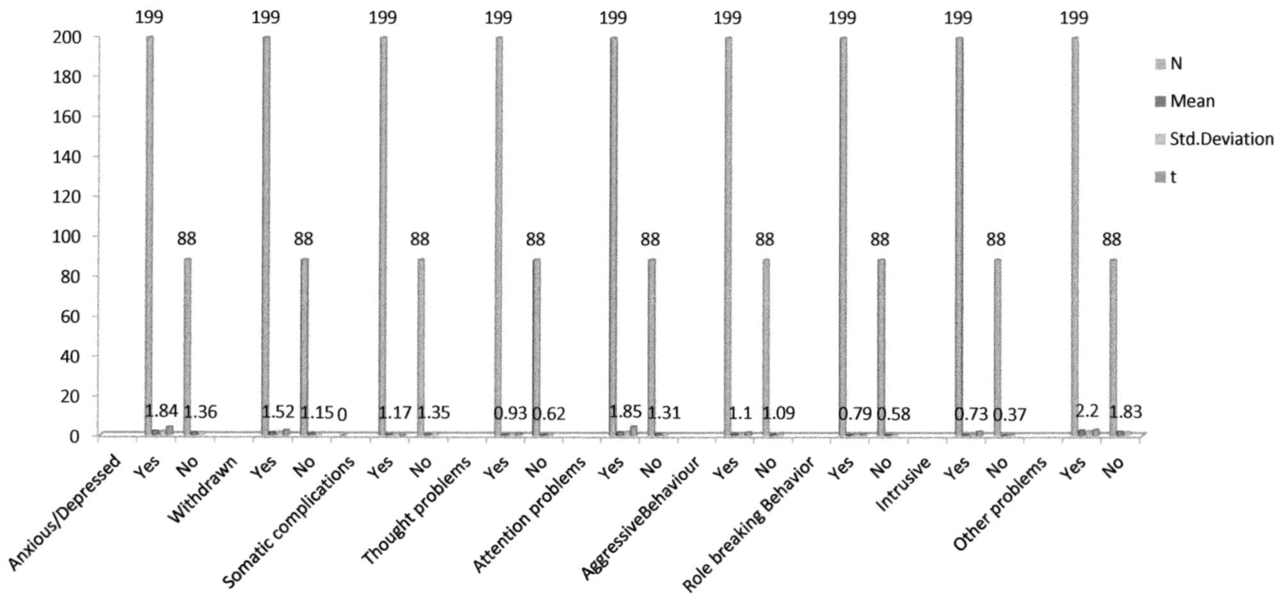

Figure 13: Social Support and Emotional & Behavioral Problems of PLHIV on Anti Retroviral Therapy.

Table-37

Social Support and Wellbeing of PLHIV on Anti Retroviral Therapy

n=287

Wellbeing subscale	Social Support	N	Mean	Std. deviation	t
Depression	Yes	199	4.83	2.689	2.513*
	No	88	3.97	2.680	
Anxiety	Yes	199	5.30	2.634	2.320*
	No	88	4.59	2.252	
Energy	Yes	199	9.32	2.293	-1.294
	No	88	9.68	2.120	
Positive wellbeing	Yes	199	9.66	3.944	-3.355**
	No	88	11.38	4.004	
Total Wellbeing	Yes	199	44.86	9.298	-3.134**
	No	88	48.50	8.972	

** $P < 0.01$ * $P < 0.05$

Table 37 shows the significant difference between social support and wellbeing. There is a significant difference found between social support and depression ($t=2.513$, $p<0.05$) anxiety ($t=2.320, p<0.05$). There is a highly significant difference found between social support and positive wellbeing ($t=3.355$, $p<0.01$), total wellbeing ($t=3.134$, $p<0.01$). This could be probably that the social support made PLHIV feels more socially safe and secure and it enhanced the feeling of self esteem resulting in feeling of wellbeing.

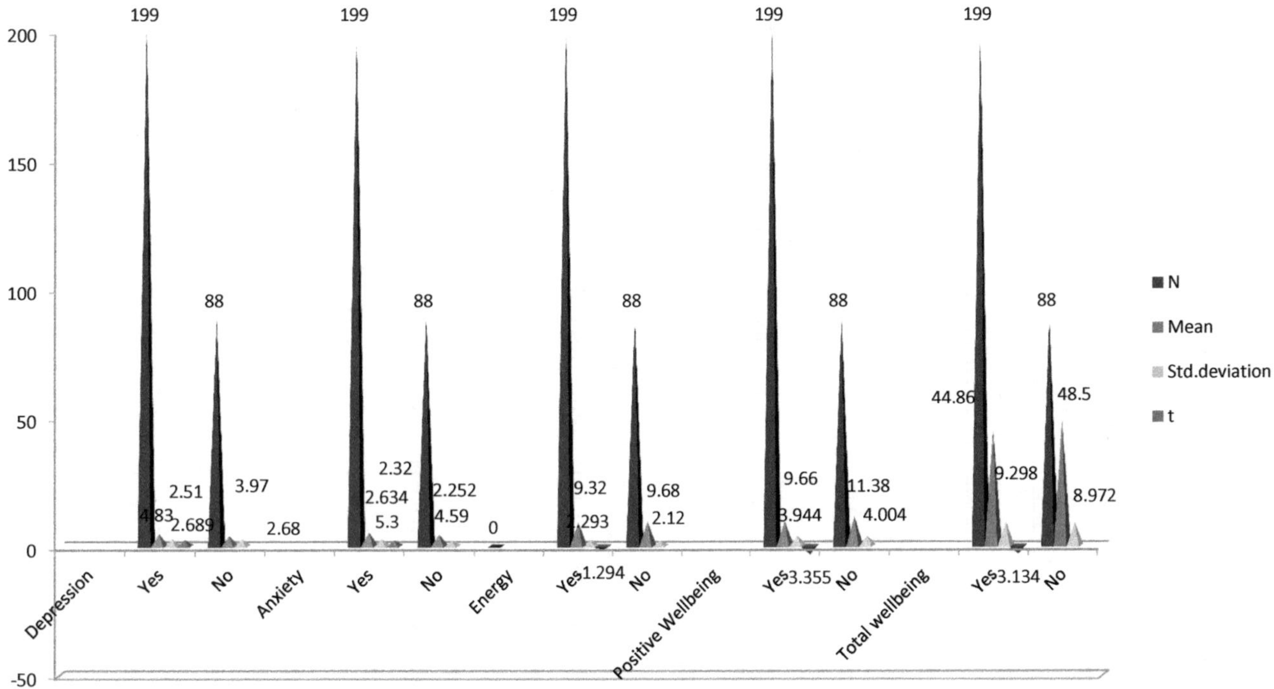

Figure 14: Social Support and Wellbeing of PLHIV on Anti Retroviral Therapy.

Table-38

Duration of ART and Emotional & Behavioral Problems of PLHIV on Anti Retroviral Therapy

n=287

Syndrome	Duration of ART	N	Mean	Std. Deviation	t
Anxious / Depressed	More than 5 Years	126	1.91	1.811	0.886
	Below 5 Years	161	1.73	1.703	
Withdrawn	More than 5 Years	126	1.25	1.390	-0.088
	Below 5 Years	161	1.26	1.469	
Somatic Complaints	More than 5 Years	126	0.56	1.395	-0.023
	Below 5 Years	161	0.56	1.089	
Thought Problems	More than 5 Years	126	0.43	0.774	0.623
	Below 5 Years	161	0.37	0.913	
Attention Problems	More than 5 Years	126	1.60	1.834	1.563
	Below 5 Years	161	1.27	1.672	
Aggressive Behavior	More than 5 Years	126	0.76	1.169	1.198
	Below 5 Years	161	0.60	1.050	
Rule Breaking Behavior	More than 5 Years	126	0.41	0.822	1.480
	Below 5 Years	161	0.28	0.663	
Intrusive	More than 5 Years	126	0.18	0.512	-0.388
	Below 5 Years	161	0.21	0.737	
Other Problems	More than 5 Years	126	2.57	2.257	0.704
	Below 5 Years	161	2.39	2.010	

** $P < 0.01$ * $P < 0.05$

Table 38 shows that there is no significant difference found between duration of ART and emotional behavioral problems.. The mean value of more than 5 years is high on anxious, thought problems, attention problems, aggressive behavior, rule breaking and other problems. The mean value of below 5 years is high on withdrawn, Somatic Complaints and intrusive behavior. It indicates, as the duration of ART is increasing, PLHIV might be feeling ART pill tiredness, side effects, fatality of a disease might be the leading factors for development of emotional behavioral problems.

Table-39

Duration of ART and Wellbeing of PLHIV on Anti Retroviral Therapy

n=287

Wellbeing subscale	Duration of ART	N	Mean	Std. deviation	t
Depression	More than 5 Years	126	4.79	2.827	1.212
	Below 5 Years	161	4.39	2.613	
Anxiety	More than 5 Years	126	5.16	2.699	0.457
	Below 5 Years	161	5.02	2.415	
Energy	More than 5 Years	126	9.64	2.076	1.432
	Below 5 Years	161	9.27	2.361	
Positive wellbeing	More than 5 Years	126	9.71	4.191	-1.750
	Below 5 Years	161	10.56	3.879	
Total wellbeing	More than 5 Years	126	45.41	9.764	-0.894
	Below 5 Years	161	46.42	8.993	

** P <0.01 * P < 0.05

Table 39 shows the significant difference between duration of ART and wellbeing. Though there is no significant association found between them, the mean value of duration of ART below 5 years is high on positive wellbeing and total wellbeing.

Table-40

HIV Disclosure status and Emotional & Behavioral Problems of PLHIV on Anti Retroviral Therapy

n=287

Syndrome	Disclosure status	N	Mean	Std. Deviation	t
Anxious / Depressed	Partner	108	0.93	1 150	-8.026**
	Family	179	2.34	1.836	
Withdrawn	Partner	108	0.78	1.202	-4.766**
	Family	179	1.54	1.485	
Somatic Complaints	Partner	108	0.37	0.816	-2.280*
	Family	179	0.67	1.413	
Thought Problems	Partner	108	0.30	0.988	-1.411
	Family	179	0.45	0.758	
Attention Problems	Partner	108	0.52	0.962	-8.4**
	Family	179	1.95	1.894	
Aggressive Behavior	Partner	108	0.36	0.912	-4.031**
	Family	179	0.86	1.170	
Rule Breaking Behavior	Partner	108	0.19	0.538	-2.855**
	Family	179	0.42	0.827	
Intrusive	Partner	108	0.16	0.763	-0.779
	Family	179	0.22	0.566	
Other problems	Partner	108	1.72	1.651	-5.192**
	Family	179	2.92	2.245	

** $P < 0.01$ * $P < 0.05$

Table 40 shows the significant difference between HIV disclosure status and emotional behavioral problems. There is a highly significant difference found between disclosure status and anxious ($t=-8.026$, $p<0.01$), withdrawn ($t=-4.766$, $p<0.01$) attention problems ($t=-8.4$, $p<0.01$), aggressive behavior ($t=-4.031$, $p<0.01$), rule breaking behavior ($t=-2.855$, $p<0.01$) and other problems ($t=-5.192$, $p<0.01$). There is a significant difference found between HIV disclosure status and somatic complaints ($t=-2.280$, $p<0.05$). The mean value of family disclosure is high on anxious, withdrawn, somatic complaints, attention problems, aggressive behavior, rule breaking behavior and other problems. The reasons for such difference could probably due to the expectations of the family from PLHIV regarding fulfillment of their needs and financial pressures and other family problems.

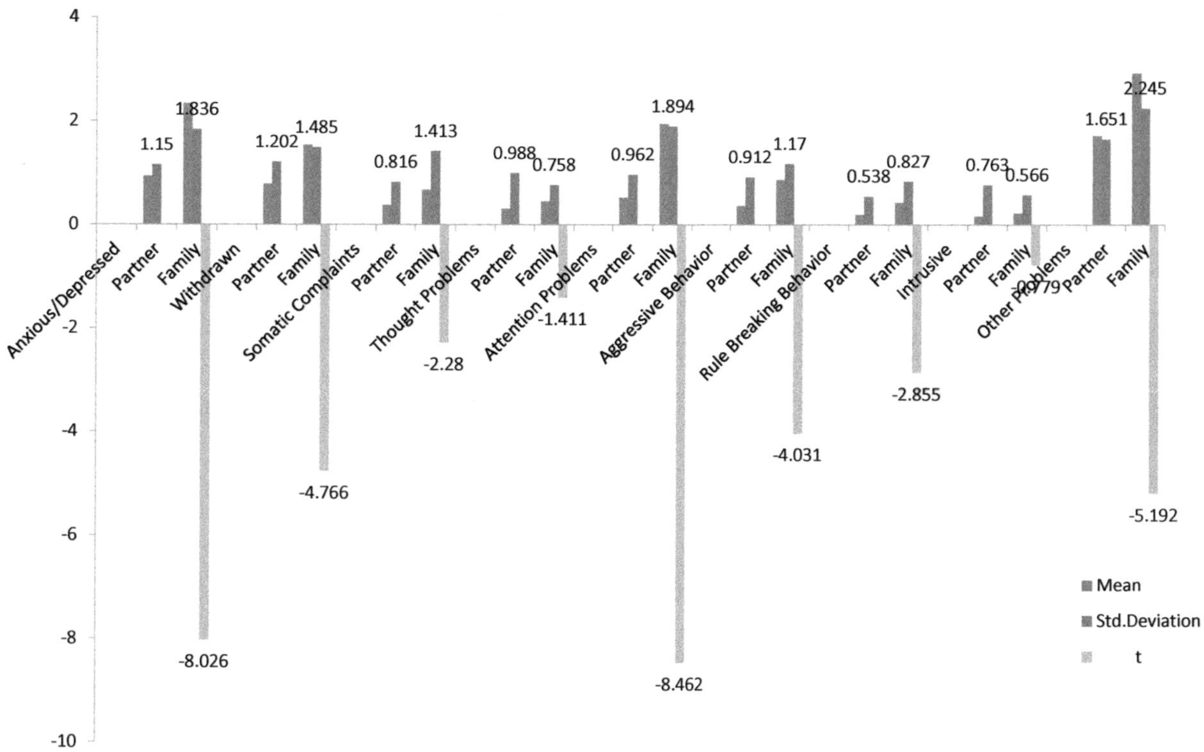

Figure 15: HIV Disclosure status and Emotional & Behavioral Problems of PLHIV on Anti Retroviral Therapy.

Table-41

HIV Disclosure status and Wellbeing of PLHIV on Anti Retroviral Therapy

n=287

Wellbeing subscale	Disclosure status	N	Mean	Std. deviation	t
Depression	Partner	108	3.73	2.433	-4.285**
	Family	179	5.07	2.753	
Anxiety	Partner	108	3.77	1.941	-7.904**
	Family	179	5.87	2.535	
Energy	Partner	108	10.20	1.807	4.999**
	Family	179	8.97	2.356	
Positive wellbeing	Partner	108	12.79	2.951	10.393**
	Family	179	8.62	3.787	
Total wellbeing	Partner	108	51.49	7.535	9.062**
	Family	179	42.65	8.737	

** $P < 0.01$ * $P < 0.05$

Table 41 shows the significant between disclosure status and wellbeing. There is a highly significant difference found between disclosure status and depression ($t=-4.285$, $p<0.01$), anxiety ($t=-7.904$, $p<0.01$), energy ($t=4.999$, $p<0.01$), positive wellbeing ($t=10.393$, $p<0.01$), total wellbeing ($t=9.062$, $p<0.01$). The mean value of family disclosure is high on depression and anxiety. This could be probably due to the pressures from the family. The mean value of partner disclosure is high on energy, positive wellbeing and total wellbeing. This indicates that the understanding between partner and PLHIV make PLHIV more energetic and feeling of wellbeing as he receives support from the partner in times of need.

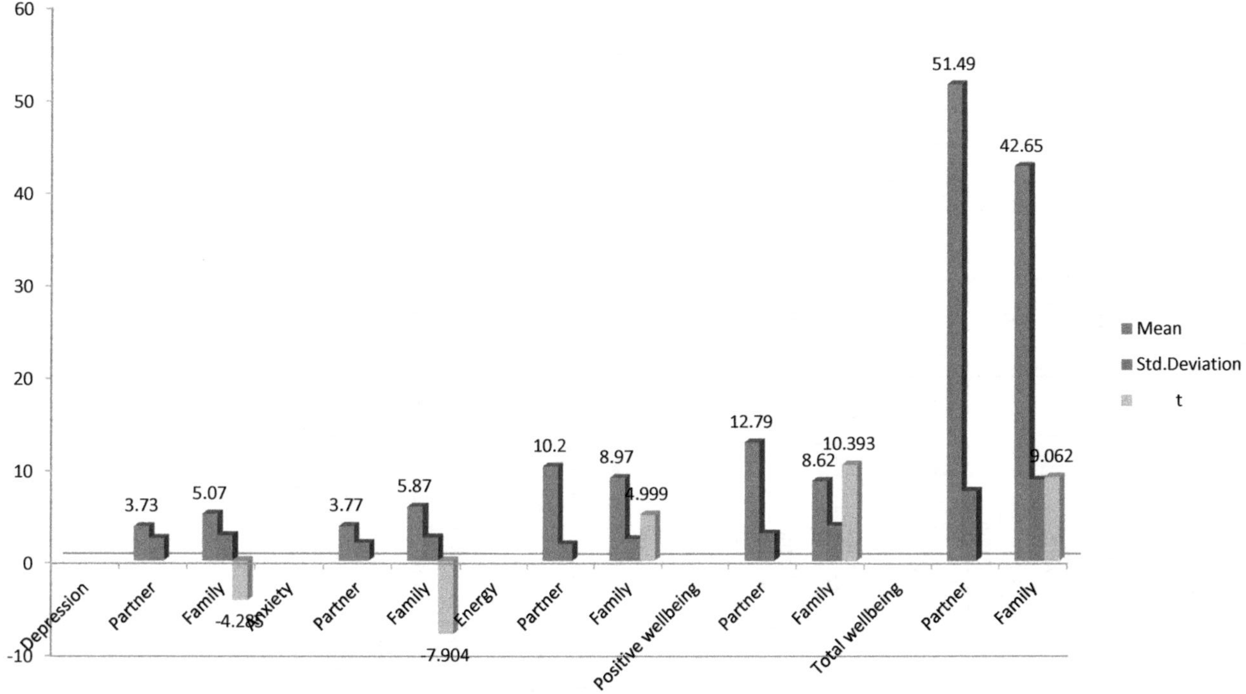

Figure 16: HIV Disclosure status and Wellbeing of PLHIV on Anti Retroviral Therapy.

Discussion

The findings of the present study reveal that there is a significant negative correlation observed between Music interest and emotional and behavioral problems. In other words, the results of the present study indicate that people who have more interest in music have lesser emotional and behavioral problems compared to those who have lesser interest in music. The mean value of low interest on music is high on depression and anxiety. In contrast, the mean value of high interest in music is found high on positive wellbeing and total wellbeing. However there are no significant differences found between music aptitude and emotional and behavioral problems as well as between music aptitude and wellbeing of PLHIV.

The present study as well as other studies thus shed light on the impact of musical interest on the wellbeing of the HIV patients. Perhaps developing interest in music could help people to improve their general wellbeing as well as reduce some of the behavioral problems. Further the findings of the present study reveal that there is no relationship between musical aptitude and emotional and behavioral problems as well as with subscales of wellbeing such as depression and anxiety etc. It is interesting to note that one may or may not have an aptitude for music, but interest in music could help a person and more specifically an HIV infected person to get over some of his behavioral and emotional problems. The review study by Archer, Bruxton & Sheffeld (2014) results supports the present study as they found improvements in anxiety, depression, coping, stress, anger and mood following psychological interventions such as art, drama, and dance and music therapy. Another study of Pothoulaki Mania et al (2012) also supports the present research findings as they found the beneficial effects of music listening on psychological variables such as anxiety, mood, relaxation and pain. Thus, the present study as well as the earlier findings suggests that creating or encouraging ones' interest in music could bring down psychological problems such as anxiety, depression, stress etc. However, further research is warranted before making conclusions in this regard.

Present study results show that there is a significant positive correlation observed between interest in music and wellbeing of PLHIV. A review of the existing literature on the music therapy also supports the findings of the present study. For instance Gallagher (2002) studied the potential for music therapy among patients with a range of chronic and or advanced illness. Physical and psychological tests were conducted both before and after music therapy. Results indicated that in addition to marked benefits in patient anxiety, mood, pain and shortness of breath, more than 80 percent of patients said their mood had improved following music therapy. Another study by Dileo & Tedaldi (2006) also supported the findings of the present study as they investigated the effects of chord arts /music therapy on physiological, psychological, social and quality of life among underserved minority HIV/AIDS patients and their care givers in Philadelphia. Participants in the program resulted in improved psychological wellbeing, reduced isolation, enhanced social support and increased hope.

Present findings also reveal that there is a significant positive correlation between emotional and behavioral problems and anxiety and depression subscales of wellbeing. In other words, the PLHIV with more emotional and behavioral problems are found with poor wellbeing as anxiety and depression seen in them. Very interestingly there is a significant negative correlation observed between emotional and behavioral problems and energy, positive wellbeing and total wellbeing. This shows that the PLHIV with less emotional and behavioral problems are found with high wellbeing. The results indicate that the PLHIV, who have emotional and behavioral problems also seem to have depression and anxiety among them. In contrast the PLHIV who have high energy, positive wellbeing and total wellbeing are found with less emotional and behavioral problems. One can be more energetic and high wellbeing if they have less emotional and behavioral problems. The present study reveals that the wellbeing status of PLHIV can be improved if emotional and behavioral problems are controlled.

The researcher also studied the significant difference between socio-demographic variables and emotional and behavioral problems and similarly between socio-demographic variables and wellbeing. There is a significant difference found between age and rule breaking behavior and no age related differences found with other emotional and behavioral problems. There is no relationship found difference found between age and wellbeing. The results suggested that adults in all age groups manifest in similar manner. The present findings support the findings of Mcgown et al (2014) who explained that the physical and psychological symptoms are common among people living with HIV at all ages and mental health is improved in old age due to greater resilience in older adults or potential for successful aging.

Regarding gender and emotional and behavioral problems and wellbeing, there is a highly significant difference found between gender and rule breaking behavior. The mean value of males is high on rule breaking behavior. A significant difference was also found between gender and depression subscale of wellbeing. The mean value of females is high on depression. Review of the existing literature Hankin & Benjamin (2001) supports the present findings as they explained the female preponderance in depression begins to emerge around age 13.They are developmentally sensitive and have elaborated cognitive vulnerability. The elaborated cognitive chain posits the negative events contribute to initial elevation of general negative affect. Generic cognitive vulnerability factors then moderate the likelihood that the initial negative affect will progress to full blown depression. These findings support the present study as the depression is found high in females. The present study findings also supporting the findinds of Liang, Lee & Farmer (2012) as their study revealed that the prevalence of depressive symptoms between serocordant and serodiscordant groups was similar. However females living with HIV reported significantly higher levels of depressive symptoms, regardless of their partner's HIV status.

Regarding occupation and emotional and behavioral problems, a highly significant difference was observed between occupation and anxious/ depressed, attention problems and a significant difference found with withdrawn behavior. The mean value of laborer is high on anxious, attention problems and withdrawn. It could be due to the loss of daily wage earning due to sickness which might have negative effect on the emotional health of PLHIV. Regarding the influence of occupation on wellbeing, there is a highly significant difference found between occupation and depression, anxiety, energy, positive wellbeing and total wellbeing. Very interestingly the mean value of laborer is high on anxiety and depression. The mean value of businessmen is high on energy, positive wellbeing and total wellbeing. Perhaps it could be due to economic security with business and flexible working hours which led to positive wellbeing. These findings substantiate the findings of Roboud & Levis (2012) as they found that employed men with secure jobs reported significantly higher mental health related quality of life than those who were non employed, but insecure employment was not associated with higher mental health scores relative to non employment.

Regarding marital status and emotional and behavioral problems, there is a highly significant difference found between marital status and thought problems, intrusiveness. This shows that the married people have more thought problems as it could be due to various pressures upon them. In contrast unmarried are found with more intrusive thoughts as it probably be due to insecurity. Further a high significant difference was found between marital status and positive wellbeing, total wellbeing and a significant difference found between marital status and wellbeing depression. The mean value of married is high on positive and total wellbeing. This could be due to the partner support enhances the wellbeing of PLHIV.

Regarding caste and emotional and behavioral problems, there is a significant difference found between caste and attention problems, other problems and no significant difference found between caste and wellbeing of PLHIV.

Regarding income and emotional and behavioral problems, there is a significant difference found between income and anxious/depression. The findings of the present study show that the low income has greater influence on the emotional health of PLHIV. Besides there is a significant difference found between income and anxiety and depression, revealing the evidence that very low income causes anxiety and depression among PLHIV. In contrast high income helps in development of positive wellbeing and total wellbeing.

Regarding number of dependents and emotional behavioral problems, there is a highly significant difference found between number of dependents and intrusive. The mean value of PLHIV of no dependents is high on intrusive behaviour compared to their counterparts. No significant difference found between wellbeing and number of dependents.

Regarding family support and emotional and behavioral problems, there is a significant difference found between family support and anxious/depressed. No significant difference found between family support and wellbeing.

Regarding social support and emotional behavioral problems, there is a significant difference found between social support and withdrawn, intrusive, other problems. Highly significant difference found with attention problems. Between social support and wellbeing, there is a significant difference found between social support and depression, anxiety and highly significant difference found between social support and positive wellbeing, total wellbeing. The present study findings supporting the findings of Kapadia et al (2016) study, whose findings have shown that the perceived receipt of positive and emotional social support as well as overall support was inversely associated with the presence of mental health problems.

Regarding HIV disclosure status and emotional behavioral problems, a highly significant difference found between HIV disclosure and anxious/depressed, withdrawn, attention problems, aggressive behavior, rule breaking behavior, other problems and a significant difference found with somatic complaints. It shows that the PLHIV who disclosed their diagnosis to family members are found with more emotional problems as it could be due to stigma and discrimination. Regarding HIV disclosure and wellbeing, there is a highly significant difference found between them. This indicates that the disclosure of HIV status to a partner helps in wellbeing of PLHIV. No significant difference in wellbeing, emotional and behavioral problems found with education, area of living, religion, living with partner and duration of ART.

The findings of the study reveals the negative correlation between interest in music and emotional behavioral problems which shows that people with music interest have lesser emotional behavioral problems. The results also show a positive relationship between interest in music and wellbeing of PLHIV. Hence the first and second hypothesis may be accepted.

The findings reveals that there is no significant relationship found between music aptitude and emotional behavioral problems. Hence the third null hypothesis which states that there will be no relationship between music aptitude and emotional behavioral problems, may be accepted. The findings also reveals that there is no significant relationship found between music aptitude and wellbeing. Hence the fourth null hypothesis which states that there will be no relationship between music aptitude and wellbeing of PLHIV, may be accepted.

The findings shows that there is no significant difference found between emotional behavioral problems and education, area of living, religion, no. of dependents, living with partner, duration of ART.As the fifth hypothesis states that there is no significant difference found between demographic variables and emotional behavioral problems, this null hypothesis may be accepted. However, there is a significant difference found between age and rule breaking behaviour, gender and rule breaking behaviour, occupation with depression, withdrawn and attention problems. Only intrusive and thought problems having significant difference with marital status. Attention problems were found a significant difference with caste. A significant difference was found between intrusive behaviour and no dependents as well as between depression and income. Also a significant difference found between family support and anxious. There was a significant difference was also seen between emotional behavioral problems and social support, HIV disclosure. Hence the above null hypothesis may be rejected with regard to certain demographic variables namely age, gender, occupation, marital status, caste, income, family support, social support and HIV disclosure.

The sixth null hypothesis states that there is no significant difference between demographic variables and wellbeing. As there is no significant difference found between wellbeing and age, education, area of living, caste, religion, no. of dependents, living with partner, family support, duration of ART, hence it may be accepted. However, there is a significant difference found between wellbeing and gender, occupation, marital status, income, social support, HIV disclosure status, it may be rejected.

CHAPTER-V

SUMMARY AND CONCLUSION

This section deals with the major findings of the study and conclusion. It also deals with the implications for practice and suggestions for future research.

Major findings of the study

- There was a high significant negative correlation found between music interest and emotional behavioral problems. No significant correlation was found between music aptitude and emotional behavioral problems.
- There was a significant positive correlation of music interest found with energy, positive wellbeing and total wellbeing of PLHIV and a significant negative correlation of music interest found with depression and anxiety.
- There was a high significant negative correlation observed between music aptitude and depression and in contrast there was a significant positive correlation observed between music aptitude and total wellbeing.
- There was a positive correlation observed between emotional behavioral problems and wellbeing subscales such as anxiety and depression. In contrast there was a significant negative correlation found between emotional behavioural problems and wellbeing subscales such as energy, positive wellbeing and total wellbeing.
- The significant difference between music interest and emotional behavioral problems, the findings revealed that the participants with low music interest were found to be high on emotional and behavioral problems and vice versa.
- There was a significant difference found between music interest and wellbeing. The mean value of high interest on music was found high with regard to wellbeing. No significant association of music aptitude found with emotional behavioral problems and wellbeing.
- The significant difference between demographical variables and emotional problems, the findings revealed that there was a significant difference found between emotional behavioral problems and age, gender, occupation, marital status, caste, income, number of dependents, family support, social support and HIV disclosure status.

- There was a significant difference found between wellbeing and gender, occupation, marital status, income, social support and HIV disclosure status. No significant difference was found with rest of the demographic variables.
- The findings revealed that interest in music has a significant influence on emotional and behavioral problems and wellbeing of PLHIV. Hence music can be suggested to PLHIV during counselling by psychologists to practice regularly to improve the psychological health of PLHIV.

Implications for Practice

The current study has implications for counseling psychologists involved in assessment and management of emotional and behavioral problems among PLHIV. Emotional and behavioral problems occurs in PLHIV due to various reasons such as fatality of disease, lack of support, repeated hospitalization due to opportunistic infections, stigma and discrimination and side effects of antiretroviral drugs. Psychologists need to consider the etiology behind the emotional and behavioral problems of PLHIV and should have a complete understanding about how it influences on mental wellbeing. Counseling psychologists should also consider the role of music therapy in prevention and management of emotional and behavioral problems and also for the promotion of mental wellbeing. Psychologists can suggest music therapy as an intervention to prevent and control emotional and behavioral problems. The findings of the present study have shed light on musical aptitude versus musical interest. While musical aptitude may or may not be present in an individual, musical interest seems to be more prevalent among many people. Accordingly a person's interest in music seem to be a better factor for suggesting one to undergo music therapy with the kind of music that a person is interested in. Thus, more than musical aptitude, musical interest seems to aid a person to reduce psychological problems such as depression, anxiety etc. and gain control over emotional and behavioral problems.

Clinicians mostly focus on diagnosis and medical interventions in control of infections associated with HIV. Mental health promotion is given less importance. Clinicians should watch for symptoms of emotional and behavioral problems such as anxiety and depression at an early stage and then they can refer them to the psychiatrist. The clinician should also focus on music therapy as an effective psychological intervention in minimizing the emotional and behavioral problems along with counseling. Psychologists and psychiatrists should consider the significance and role of music therapy besides giving anti-psychiatric drugs.

Although no statistically significant demographic related association of emotional and behavioral problems, wellbeing found with education, religion, area of living, living with partner, duration of ART. This lack of significant findings doesn't mean that no significant difference exists. Psychologists should also consider the influence of socio-demographic variables on emotional behavioral problems and wellbeing during assessment.

Counseling psychologists like many other health care professionals, should be encouraged to update their knowledge about developments in their field by attending various trainings and conferences. Ideally professionals from different disciplines such as health care providers, social workers, psychologists and other related professionals such as music therapists should work collaboratively to provide comprehensive support to adults affected by HIV/AIDS worldwide.

Suggestions for future Research

The current study has implications for future research. The current study assessed the music appreciation of PLHIV through music appreciation questionnaire and adopted descriptive research design. If the researcher provides a music therapy to PLHIV of experimental group and withholding it for control group and observe the effect of music therapy through adopting experimental design and finds the relationship between music therapy and emotional behavioral problems would be yielding better results than descriptive study. Thus it is suggested that future researcher can be conducted as experimental study with a provision of music therapy to PLHIV.

Counseling psychologists who conduct research should aim to have a diverse sample that can be generalized to the population. The current study population consisted of PLHIV attending two major ART centers at two major cities, Visakhapatnam in Andhra Pradesh, and Hyderabad at Telangana. In addition the sample size was not too big. Thus it is suggested that future researchers aim to have a bigger and more diverse sample size from wider area.

In the current study, the researcher educated the PLHIV on importance of prevention of re infection, adherence to ART and adoption of healthy lifestyle, but a planned counseling process was not followed. Thus it is suggested that future researchers can conduct counseling and if they study the effect of counseling on control of emotional and behavioral problems, the study would yield results in specific direction.

Additionally, there were some instrument limitations in the current study. If this study were to the replicated, it is suggested that clinicians implement not only more and valid measures, but also numerous instruments to measure the dependent variables. The current study used an instrument to measure the various syndromes of emotional behavioral problems. If a separate instrument is used to measure the emotional problems and separately the behavioral problems, it would have yielded the better results. Therefore, for a replication of the study, it suggests that various instruments can be used. Lastly, rather than using a parent rating scale, researchers may want to include a self-report measure to investigate emotional characteristics and behavioral characteristics.

Limitations

1. Data were obtained from a limited sample. Further the study was limited to the ART centers of two hospitals only.
2. Random sampling method could not be used by the investigator.

Conclusion

Four main findings emerged from the present study. First, results indicated that there is a significant negative correlation observed between interest in music and emotional and behavioral problems. So PLHIV who had more interest in music were found with less emotional and behavioral problems. Second, results from this study reveal that there is a significant positive correlation found between interest in music and wellbeing of PLHIV. Wellbeing was found high among PLHIV who had more interest in music. Third, results from this study indicate that there is a significant positive correlation found between emotional and behavioral problems and subscales of wellbeing such as anxiety, depression and a significant negative correlation found between emotional and behavioral problems and energy, positive wellbeing and total wellbeing of PLHIV on Antiretroviral Therapy. PLHIV who had more emotional and behavioral problems were found with depression and anxiety. In contrast, wellbeing was found high among PLHIV who had less emotional and behavioral problems.

Fourth, there is a significant difference between emotional behavioral problems and age, gender, occupation, marital status, caste, income, number of dependents, family support, social support, HIV disclosure status. Rule breaking behaviour was found high in 30 to 40 yrs of age group and it was also found high in males. Anxious/depression, attention problems were found more in labourers. Thought problems and intrusive behaviour were found high in single women than married.

Depression was found high in below Rs 3000 per month income group. Intrusive behaviour found high in PLHIV with no dependents. Depression was found high in PLHIV having family support. Withdrawn, attention problems, intrusive behavior were seen more in PLHIV having social support. Depression, withdrawn, somatic problems, attention problems and rule breaking behaviour found high in PLHIV who had disclosed their HIV status to family members than disclosed to partner.

Further there is a significant difference found between wellbeing and gender, occupation, marital status, income, social support and HIV disclosure status. Depression was found more in females. More energy and positive wellbeing seen in business men and least energy, least positive wellbeing found in labourers. Positive wellbeing and total wellbeing was found high in married. Depression was found high in less income (below Rs 3000 per month) group and less depression seen in high income group. Positive wellbeing seen high in high income group. Depression was found high in PLHIV having family support.. Positive wellbeing, total wellbeing were seen less in PLHIV having social support. Depression and anxiety was found high in PLHIV having social support. Total wellbeing was found high in PLHIV having disclosed their HIV status to only partner than disclosed to family members. No significant difference was found with rest of the variables.

In conclusion, it is revealed through this study that musical interest has a significant influence on the wellbeing and emotional and behavioral dimensions of the health of PLHIV. Hence music can be suggested to PLHIV during counselling as part of their life by psychologists to improve the wellbeing of PLHIV.

CPSIA information can be obtained
at www.ICGtesting.com
Printed in the USA
LVHW021316031222
734485LV00013B/1030

9 781805 452294